THE AESTHETICS OF ENGINEERING DESIGN

the aesthetics of engineering design

Fred Ashford

Hon Des RCA FSIA FRSA

bb
BUSINESS BOOKS LIMITED
London

First published 1969

© FRED C. ASHFORD, 1969

All rights reserved. Except for normal review purposes, no part of this book may be reproduced or utilized in any form or by any means, electronic or mechanical, including photocopying, recording, or by any information storage and retrieval system, without permission of the publishers

S.B.N. 220.79855.9

This book has been set in 11 pt. Pilgrim printed in Great Britain by Hazell Watson & Viney Ltd, Aylesbury, Bucks for the publishers, Business Books Limited (registered office: 180 Fleet Street, London, E.C.4); publishing office: Mercury House, Waterloo Road, London, S.E.1

MADE AND PRINTED IN GREAT BRITAIN

contents

1 INTRODUCTION — 1

2 THE FUNCTION OF AESTHETICS IN ENGINEERING DESIGN — 11
    Ergonomics — 15
    The commercial value of aesthetics — 16
    Rationalization and optimization — 24

3 PERCEPTION — 30
    Irrationality — 33
    Basic visual elements — 39
    Visual competition — 44
    Sense-impressions — 46

4 FORM — 51
    Composition — 51
    Surface — 53
    Visual unity — 55
    Proportion — 58
    Harmonic relationship — 64
    Secondary forms — 66
    Direction — 70

5 AESTHETIC MANOEUVRE — 79
    Clarity of form — 82
    Clarity of expression — 85
    Balance — 87
    Symmetric or asymmetric form? — 90

6 SURFACE TREATMENT — 92
    Letters and numerals — 92
    Layout and balance — 100

CONTENTS

Nicety of display 104
Size 106
Colour 108
Aesthetic aims and the influence of fashion 114
Determination of product quality 117

GLOSSARY OF AESTHETIC TERMS 121

INDEX 127

ONE

introduction

DO NOT think that this is yet another book about industrial design. If possible, erase that unfortunate phrase from your mind for, like all expedients or temporary solutions, its usefulness is proscribed by the very factors that generated it. The phrase 'industrial design' is unfortunate in that national bodies and senior establishments engaged in engineering design education are tempted by it to isolate completely the human aspects from mechanical and structural design. They lump them together as 'industrial design' and offer them as but a shallow 'appreciation' of something they know to be very important, but the responsibility of someone other than the engineering designer. Thus, the very people who should be receiving enlightenment are denied it and led to believe, like their tutors, that the human qualities of engineering design are not their responsibility. For so long as this unfortunate phrase endures, we shall have the present artificial, illogical situation in which industry's designers are inadequately educated and will go on producing problems for aesthetics and ergonomic specialists to clear up, instead of producing work of a higher quality enabling the specialists to make their full contribution. Devoting considerable skill and effort merely to make the best of bad jobs is scarcely an intelligent way of carrying on. Industrial design was a convenient phrase to imply reference to the appearance, the convenience and the rationale of function and manufacture of products, qualities which had come to be neglected. Some external influence, outside of the many industries, had become necessary in order to ensure that this neglect was corrected. This new activity did make a most valuable contribution to commerce and industry. It still does so but in a steadily decreasing number of instances because it is a self-eliminating activity. What endure are the qualities to which it drew attention, the normal and natural aspects of

2 THE AESTHETICS OF ENGINEERING DESIGN

industry which had been allowed to become neglected. Many of the earlier advantages conferred by industrial design have become specializations in their own right, such as ergonomics, value engineering, product rationalization and corporate identity, while better realization of the potential importance of industry's own designers has grown. That these designers have, on the whole, neglected certain aspects of their work and have been by-passed or not catered for by our educational authorities, makes some form of self-help a necessity for the majority. To assist in this is the aim of this book. It is aimed at engineering designers because engineering methods apply to a greater number of industries than do the methods of any other activity, and aesthetics constitutes a fundamental part of engineering.

Engineering is an art, with a history as long as that of any other art and with an equally illustrious roll of practitioners. In 2283 B.C. a Chinese engineer known as The Great Yu designed and carried out flood control and irrigation works which literally saved his country after an unprecedented deluge. As a tribute to engineering he was made Emperor upon the death of the current Emperor, Shun. About 500 B.C. Greek and Persian engineers were producing such notable achievements as the bridge of boats across the Bosphorus and a 3000 foot long water tunnel driven simultaneously from both ends. About 200 B.C. the mathematician and engineer, Archimedes, designed and supervised the construction of a number of war machines which checked the Roman invasion of South East Europe. He was later killed by a Roman soldier, and Cicero directed that a cylinder and a sphere be placed upon his grave to signify that he was the first man to calculate the volumes of those two solids. About A.D. 1500–50 Agricola did great work in engineering, notably in the field of metallurgy, leaving behind him a series of twelve well-documented books, rather like Leonardo da Vinci's famous notebooks but without their graphic richness.

Engineering differs from some other arts in that it is always objective and always motivated by a desire that it should be useful to mankind, unlike painting and sculpture which are entirely subjective and not done with any conscious desire that they should be utilitarian. That this virtue is sometimes overlaid by a desire for personal profit does not invalidate it. It was no doubt due only to the fact that engineering had inevitably to be associated with a great deal of noise, mess and crude basic processes that it did not become established, as did architecture, as the Mother of the Arts. With architecture nowadays so dependent upon and conditioned by engineering-based technologies, this is today a somewhat hollow claim. It is significant to note that the Greek word 'architect' was not limited to those concerned only with buildings but with all branches of technology in use at the time. An eighteenth-century German engineer was known as a Master of Arts; even today engineers in that country are still known by the same title of *Kunstmeister*. Goethe held that the engineer and the technologist contribute something more than their material structures to society, that they contribute a philosophy. The purpose of this overture is to restate that engineering is a creative art; that engineering design, the very part of the total activity in which invention, innovation and imagination all play a part, should reflect a philosophy and a purpose as commendable as that of any other profession. It is merely unfortunate that, just as the prestige of engineering has been permitted to decline and the image of the engineer to become that of a rather uncouth fellow in oily overalls, carrying a large spanner, so the proliferation of detailed drawings, with a consequent growth of non-creative draughting, has stigmatized drawing offices and the graphic expression of engineering thought generally.

Following the tendency towards specialization, the human aspects of engineering design, principally the aesthetic and ergonomic, have somewhat regrettably become more the concern of others than of the engineering designer himself. Regrettably, because with engineering, more than with any other artefact, aesthetic and ergonomic quality is inseparable from functional and material decisions. Any belief that aspects of the subject which are very much the concern of everybody in an engineering design team, can be 'hived-off' and dealt with by specialists, who by definition are not practitioners of engineering, is clearly due to unsound thinking. It is quite true that overall co-ordination by a specialist is possible. It can however only be really successful where other members of the design team are not all the while making decisions which can adversely affect and prejudice the very qualities which it is desired to establish. The purpose of this book is to encourage engineers to take a broad, complete view of their subject, and to ensure proper recognition of those aspects of it which have tended, quite artificially, to become the responsibility of someone else. Those aspects which have tended to become lumped together under the designation of 'industrial design' are, of course, normal and natural aspects of engineering design, and they have been so since time immemorial. This book is also an endeavour to fill in a curious gap in the bibliography of engineering and aesthetics, a gap which has encouraged the establishment of the gulf which certainly often exists between aesthetes and practical men.

From the commencement of engineering as a recognized activity, aesthetics has been an inseparable part of it. Leonardo da Vinci, like so many other creative men in history, did a spell as a military engineer. His famous notebooks are full of sketches of hypothetical rather than actual engineering projects but these, nevertheless, firmly establish an association between art and engineering, between aesthetics and nuts and bolts. The frequent citing of Leonardo as archetype for the artist-engineer may induce a tendency to boredom and cynicism, but for engineers wishing to explore the aesthetic aspects of their work, Leonardo's principles and methods hold the greatest value. He was not an engineer in the sense that we understand the term today; he was the *uomo universale*, an inspired and sensitive artist-artisan, for which in these times of wide dissemination of knowledge and of executive specialization there may be no real equivalent. Beyond the possession of a highly enquiring mind, an extraordinary visionary capacity and the great graphic ability which enabled him to present his engineering designs to the best advantage, he was not qualified in the currently understood sense. Few of his designs had to stand the acid test of realization in hardware; no doubt very few of them would have stood that test, not because their engineering was wrong in principle but because, in addition to giving pictorial expression for the first time to some existing inventions, many of Leonardo's own inventions were centuries ahead of the technological possibility of their realization. He was, it seems, more interested in an idea than in the possibility of carrying it out. He spent his declining years in France, and when he died, in a house on that pleasant little hill in Amboise overlooking the River Loire, record has it that he was a disillusioned, unfulfilled man. He was a bastard, almost certainly a homosexual, demonstrably hermaphroditic in outlook, and it is conceivable that he was subject to considerable inner tensions and stress. So what kept him going to produce the magnificent volume of achievement credited to him? Apart from his obvious executive skills, in what did his conceptual strength lie? It almost certainly lay in the unwavering belief that observed and understood truth is the only sure basis for any creative activity. He discounted fashions, conventions and formulae, and he was always going back to square one which, for him, was Nature. This prime source has inspired many great engineers before and since Leonardo.

Nasmyth was a great lover of nature; Maudslay laid great stress upon avoiding sharp angles and corners in the design of metal objects, pointing to the junction of the fingers to the palm of the hand and to the branches of a tree to the trunk as examples of design perfectly suited to function and to material. Square one for engineers seeking an understanding of the aesthetics of man-made objects can also be an appreciation of those objects themselves, of any object. What function is it intended to fulfil? Why is it shaped as it is? How well does the shape suit the function? What has determined the material and is the shape suited to the working of that material? Is perception of the object easy, or do any features interfere with total perception? What is the eye having to do, and what is the brain making of what the eye sees? Leonardo called experience the Mother of all Arts. Certainly the personal discovery of truth always results in a surer understanding of it than does any revelation of it by a teacher, however good he may be. Personal discovery may take a little longer. The aim of this book is to indicate the road to it, with illuminated signposts which may make the going more rapid.

Expectation of aesthetic quality in engineering design was satisfied by the work of many engineers. Unfulfilment was ushered in by the shift of emphasis from quality to quantity as a direct result of the Industrial Revolution. This unfulfilment has been maintained by the concentration on specialized technical function and by the splintering of the engineering activity, a condition aggravated by the omission from the education and training of engineers of any recognition of the human aspects of their work. Recently much activity and concern has been associated with the restoration of those aspects, notably the aesthetic and ergonomic. Current provisions for ensuring the former are not so good as

6 THE AESTHETICS OF ENGINEERING DESIGN

for the latter, and provision for either is minimal, hardly scratching at the surface. Until there has been a complete reappraisal of arrangements for ensuring human qualities in engineering the situation remains very much one of self-help. An exposition of aesthetics in language that engineers can understand, in the analogy of their own work, may assist this process. Aesthetic quality in engineering can be achieved in two ways: through elegance of concept and through elegance of realization. When both are achieved we have completely satisfying objects, but these are all too rare. Too often we have elegant engineering thought presented in the form of crude hardware, robbing it of much of its potential, or we have great elegance of realization applied to trite,

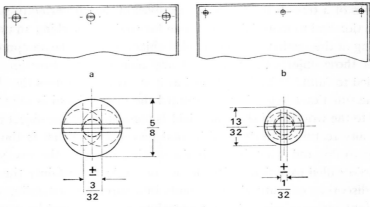

Fig. 1.1

indifferent thinking. The use of the word 'elegant' here should not be misunderstood. To many it means only great refinement or luxury, to most it means graceful or tasteful, but it can also imply utmost simplicity and economy. With concepts it can describe any perfect, Platonic solution to a problem which may be a mathematical one or one connected with any worldly or material activity. With percepts it can describe something realized in hardware with the greatest visual simplicity and the greatest economy of material and effort. Good appearance is by no means determined only by the designers; elegance in realization is often the result of first-class planning and execution. An example of how a considerable improvement in appearance was achieved by improved standards of workmanship is illustrated, Fig. 1.1. This simple example could be repeated in connection with many methods of securing or finishing parts.

The sheet metal fabrications of a company were marred by very large and conspicuous fastenings, made necessary by poor standards of hole placing and part folding and bending. To be able to insert fastenings into misaligned holes it was necessary to make the holes much larger than the diameter of the fastening and the head of the fastening had to be large enough to cover the hole. By improving standards of hole positioning and of part forming, general tolerances were reduced to about one-third of what they were, and the use of less conspicuous fastenings of almost half the former head size became possible. Over the last thirty years efforts have been made to educate some artists in such a way that they might acquire an appreciation of the principles, processes and disciplines of engineering. While this has undoubtedly made some contribution to retaining an element of creativity in engineering design it remains an unsatisfactory, artificial solution. Where deficiency or sickness exists, straining one's resources to produce external remedies is no real solution. How much better it is to concentrate upon building up the health of the body so that its need of external remedies ceases! Until we get back to the truth that aesthetic quality is an essential part of engineering design, that it can often be of greater commercial significance than many material factors, and that all those engaged in engineering design and development must have some appreciation of it, engineering design will not achieve the results it should. To consider aesthetic quality as but a refinement, a luxury that cannot always be afforded and in any case the concern of some specialist, is completely wrong thinking.

Engineering is concerned with the actuality of round, three-dimensional form, solid hardware; the examination of aesthetic cause and effect in two dimensions may at first therefore seem to have little relevance. It may not seem so improbable, however, if it is remembered that retinal images are two-dimensional in that they are on one plane; that depth perception, or stereoscopic vision, is the result of an elaborate computation by the brain of the disparity between each retinal image, between what is seen by each eye. Provided that there is a capacity to think of objects in terms of three dimensions and not only of two, the resolution of the principles of aesthetic cause and effect can quite conveniently be carried out upon the two-dimensional page of a book. In committing oneself to the limitation of a subject which a book imposes there is always the danger that parts of the content may be taken and seen as rather more of the whole than they actually are. This hazard is particularly present when the subject of the book is aesthetics, without rules, values of formulae, even without a language of its own. Many considerations need to be included in order to convey the idea of a visual mode of thought, and there is always a danger of any one of a number of basic aesthetic aims being interpreted as an end in itself. Mention of just one of a number of ideal solutions is apt to get oneself labelled as a dedicated promoter of that one solution. To avoid such a possibility and to show how wrong such an assumption could be, let us look at a possible example of how this could happen.

a   b   c
Fig. 1.2

Taken to an apparently logical conclusion, the ultimate, most aesthetically satisfying form could be a grey, circular disc shape, Fig. 1.2a, a form that is simple, continuous and perhaps the easiest of all to perceive. The colour, grey, is neutral, achromatic; it avoids emotional decisions about hue; in addition, it avoids sharp contrast with the surround which could impose greater activity upon the eye. Let us imagine that such a conclusion was arrived at, and see what would inevitably happen. To satisfy a fundamental human need for variety the circular form would no doubt quickly become an elliptical one, Fig. 1.2b. An ellipse, however, might not be suited to some functional purposes of the object to which the form might be applied, nor to some methods of making it, and we could expect variations of form taking us farther and farther away from the original circular form, Fig. 1.2c. To satisfy the need for variety, and to integrate the form more suitably into a number of normal backgrounds, the grey tone would almost certainly be replaced by a variety of colour. If at this stage someone were to protest and say 'Hey, what about that optimum aesthetic form, why have we departed so far from it?' the answers he would receive would reveal the real reasons for most things being designed as they are. They would show how, instead of trying to impose a theoretically ideal aesthetic quality upon a situation, it is better to integrate it with the situation. In short, solutions must be adapted to facts; it is usually impossible to adapt the facts to the solution. Nothing that is written here should be taken as hard-and-fast rules or as dogmatic opinion, but only as general and, it is hoped, helpful considerations of the presence and the possibilities of something as sensitive and as intangible as aesthetic quality in the mundane connection of nuts and bolts and rough, tough and often crude industrial processes.

Realization that analogous qualities must be inferred merely to provide a basis upon which to discuss or to manipulate aesthetic values can be an opening-up of understanding of a subject that has come too much to be regarded as being esoteric and too far removed from the hardware situation by which it is generated. The aim of this exposition is to examine some basic principles and effects in terms of the analogies the fine-art world is obliged to use, to explain such terminology of value and manoeuvre as are normally used to discuss aesthetics and to try to show how and why they have been evolved. A glossary of terms which appears on page 121 may thus assist understanding, as aesthetics has no language of its own and has to borrow phrases and words used in other connections so giving them a duality of meaning. For example, an artist referring to the placing of one form upon another may make reference to it sliding off or breaking out, while the engineer, who knows very well that the forms under discussion are solid metal castings or strong weldments, does not understand what on earth the artist is talking about. He does not understand that the terminology used is with reference to what appears to be happening or just about to happen visually. Reference to the factor of irrationality in perception should reassure the engineer of the validity of the artist's observations, while reference to the glossary of terms should familiarize him with some of the jargon of visual technique.

Engineers and artists of the same nation may speak the same language, yet to each some identical words may have different meanings; no mere aesthetic caprice, but due to unavoidable verbal analogy leading to the attachment of new meanings and shades of meanings to some words. The glossary contains the most generally used words and an appreciation of their meaning in the aesthetic context should assist communication between engineers and artists.

TWO

the function of aesthetics in engineering design

A CARDINAL principle to be observed, ironically the one most responsible for the neglect of human qualities in engineering, is prior recognition of engineering function. Great success in the achievement of human qualities is of little avail if the product fails to do what it was designed to do and to do it well. On the other hand, however well a product achieves its mechanical and material function, its overall success is dependent upon how acceptable it is to human beings. Only complete acceptance ensures the realization of the maximum engineering potential, while incomplete acceptance can only ensure something less than maximum potential. These are just simple, hard facts.

All engineering products are articles of utility. The prime design consideration must be that they will function correctly and be constructed soundly enough to go on functioning for the extent of their planned service-life. A clock, a refrigerator and a machine tool are expected, above everything else, to keep accurate time, to maintain low temperature efficiently and to provide accuracy, efficiency and dependability. However beautiful or convenient to handle they may be, they would be of reduced value, or of no value at all, if they failed to meet the prime requirement of satisfactory mechanical and structural function. This must be the responsibility of the engineer. In discharging it he cannot avoid making decisions which must affect the form, arrangement and other aspects of the aesthetic quality of the object. Given the satisfaction of the prime requirement, the total usefulness of a product is greatest where the apparance and the amenity make it so acceptable that the maximum realization of any mechanical or structural virtue is guaranteed. This happy state of affairs can be achieved with the aid of aesthetic and ergonomic specialists, although they must enjoy the enlightened support of the rest of the engineering design team if they are to be successful. Better still, it could be achieved by the total engineering design effort of the engineers themselves. There is little doubt that given adequate appreciation of aesthetic and ergonomic quality on the part of the engineering designers a great many engineering products could be brought to a much higher level of all-round quality than they normally are. Decisions about mechanical and structural function must have priority over appearance. It is therefore essential for all engineering designers to be aware of the aesthetic implications of their activities. The specialist-team concept is not a valid one so far as the really formative stages of engineering design are concerned. It can only work where design is rigidly specified and tightly cross-referenced, by which time basic aesthetic damage has nearly always been done, and it is too late for fundamental change. The basic philosophy of this work is not therefore to disparage the efforts of the aesthetic and ergonomic specialists, who make a valuable contribution, often in the face of great practical difficulties, but to suggest how much better it would be if engineering designers were themselves able to ensure high quality in every aspect of their products.

The opportunities for achieving aesthetic quality in engineering design are considerable. Leaving aside for the moment the aesthetic quality of elegance of thought in a concept, and considering only the wealth of possibility in its realization in hardware, there is an almost infinite variety of materials capable of different shape-characteristics arising from appropriate forms and from alternative forming processes. There is also the great variety of different surface treatments; generally applied finishes such as paint, metal and plastic films or the rearrangement of the surface film by chemical means. To this material aesthetic can be added the aesthetic of shape, form, arrangement and the achievement of an overall product-quality or characteristic. How engineering ever came to be regarded as a dull, plebeian activity it is difficult to understand, but it has; engineering itself is no doubt most to blame, having for so long neglected the human aspects. There is not only value in the recognition of these qualities; there is a great deal of satisfaction in the apt selection of correct visual qualities from the apparent abundance. It must at the same time be realized that structural, mechanical and economic considerations can sometimes narrow the field to a choice containing a measure of some aesthetic incompatibility. Where by good selection an end-result is achieved which meets all requirements, including, of course, the aesthetic ones, the satisfaction can be equal to that of any other creative activity. Not only for the painter and the sculptor is there the joy of creation. A good engineering product capable of bringing pleasure to many others can be equally satisfying. One of the functions of aesthetics in engineering design is to indicate function and purpose. Simple, easy-to-perceive forms which recognize function and purpose as two things conditioning their design automatically do this. Hence the emphasis throughout this book upon that kind of form. Having rid engineering of its early classical architectural disguise and, more recently, made it ashamed of spurious aerodynamicism, it is particularly important to avoid any silly, self-conscious styling to indicate purpose and function. An engineering aim properly analysed should control the basic design sufficiently to ensure adequate indication and to make necessary only a minimum of cosmetic addition, consisting possibly of the display of any necessary information and local emphasis by colour. Basic design failing to indicate function and purpose is somehow wrong and is usually due to having rated manufacturing convenience too highly and at the expense of everything else. Forms which do not indicate their function and purpose clearly usually do not fulfil them very well and forms disguised as, or even looking like, other things are at a disadvantage perceptually. So here we are, back to square one: good basic design is the only acceptable foundation and this is why aesthetic quality cannot be left, like icing to be added to a cake. The cake itself must be right and the selection, mixing and baking of the ingredients are the responsibility of the cook, who, in the engineering context, is the engineering designer. It might be queried, since eye and brain are capable of perceiving the most complex visual arrangements, what is the purpose of the pursuit of simple forms and straightforward visual arrangements? Is there not a danger of the visual world becoming too simplified, stark, plain and bereft of intricate visual detail and richness? No, certainly not. Most engineering forms are associated with objects of utility and connected with functions and activities demanding a measure of attention, while few are associated with situations of complete relaxation and leisure. That is why the intention is generally that the nature and purpose of the object shall be readily identifiable and why there is rarely any intention that the object should be visually an end in itself. Engineering objects are almost always components of systems for providing some useful service and instant recognition is necessary to facilitate their use. Any arrangement of form, colour or decoration to combine sensuous pleasure must be a secondary aim, but provided that the primary aim is achieved there is nothing to exclude achievement of the secondary aim and the possibility of an over-clinical environment is not great. There is usually a natural restraining influence exerted by the primary aim over the secondary aim, and the creation of a visually over-rich, too intensely-absorbing engineering object is a clear signal to be read as 'Back to square one' or, more colloquially, 'Back to the old drawing board'.

To realize the functions of aesthetics in engineering design properly it is, of course, necessary to be able to think visually, to be able to form percepts as well as concepts. The development of an ability to think thus is like learning the grammar of a language as well as the full meaning of the separate words. Without knowing the grammar it may certainly be possible to identify most of the words, yet impossible to assemble them in the most meaningful way, or even to give them any real meaning at all. Once the grammar is learnt, however, there is no situation for which the right words cannot be found and arranged in the best way. Because the variations of sensory stimuli are almost infinite there can be no simple hard-and-fast rules with aesthetics, which disappoints many who expect aesthetic sensibility to be a straightforward, cut-and-dried matter requiring only decent observance of established disciplines. The truth is that it is not too easy to start with, as learning a grammar calls for greater effort than would the observance of simple rules, but having learnt it, the understanding can be applied to an infinite number of situations quite naturally and without much further effort. Immediate apprehension without having to reason amounts to intuition; if intuition as a basis is coupled with reason and logic there is really no sense-situation that cannot be adequately met.

Another function of aesthetics in engineering design is the satisfaction of human requirements. That is a neat phrase, much bandied-about, but compared with ensuring that an engineering product will work well and be constructed well enough to keep on working, how important to engineers is the satisfaction of human requirements? In the long term it is of the greatest importance for, although in established communities basic needs are adequately satisfied, preference is increasingly shown to products seeming to have been tailored to suit the individual. Note the qualification, seeming, because personal tailoring and off-the-peg availability are incompatible, so how can the virtues of the one be brought to the requirements of the other? The answer is through the creation of an empathy, through suiting the product dimensionally to a suitable percentage of human users with adjustments for personal fitting: through making the product sympathetic in texture and colour and ensuring that its arrangement makes its use an agreeable experience. There is no mystery about any of these things; the knowledge and skills to ensure that they are realized are available, but it is essential that the engineering design is carried out in such a way that these human qualities are properly integrated rather than having to be applied, which can never be wholly successful. It should be clear that without the recognition of these human qualities by the engineering designer, any specialists can only be making the best of a bad job, working from the outside inwards and coming up against stop after stop.

ergonomics

Aesthetic and ergonomic aims are, for the most part, synonymous and if aesthetic decision is accepted as being the result of objectivity rather than of arbitrary selection, it could be argued that aesthetics could be adequately covered by a good book on ergonomics. This is not the case, although the existence of a number of good books on ergonomics has made it unnecessary to repeat the charts, diagrams, data and exposition of principles to be found in them. Further, in this book, it is assumed at all times that the human aspects of design, the fitting of what is being designed to the human user or operator, are recognized and are influencing the design. It was not considered essential to have a separate section dealing with ergonomics and recognition of ergonomic requirements should be apparent throughout the separate sections of the book.

As shaping and placing must clearly be very much a part of aesthetic quality, it is worth while examining the difference between ergonomics and aesthetics. Ergonomic disciplines are related to human beings who are variables and, to some extent, unpredictable and, although they are not so finite as engineering disciplines which are related to controlled, inert materials and to calculable effects, they are more rigid and more finite than aesthetic disciplines. They determine the position of points in space, but not the linking of those points nor the visual characteristics of the resulting constructions. Good ergonomic quality can be assured in several ways: by reference to specialists; reference to the considerable amount of tabulated data now available; by full-scale layouts; full-size sketching on a gridded area of the drawing office wall and by wood-lath and hardboard, card and plasticine mock-ups. Travesties of ergonomic requirements perpetrated in the interests of manufacturing convenience, mental laziness or indifference are no more than nails in manufacturers' coffins.

Although this section on ergonomics is limited for the reasons stated, it is more than a token recognition of a most important aspect of engineering design. Just as material function is a prime consideration, so is the correct scaling, shaping and positioning of all points of contact between an engineering product and a human being. To avoid the unpractical splintering of engineering design into multiple specializations, already manifest with aesthetics, engineering designers must themselves make sympathetic recognition of ergonomic requirements, but this view does not preclude consultation with specialists when desired. A little more effort is being made to inject an ergonomic content into engineering design education than is being made to inject any worthwhile aesthetic content, but it is not yet enough and it must grow from being a separate subject to something which naturally permeates every separate aspect of engineering design.

the commercial value of aesthetics

That aesthetic quality is able to make engineering products more readily acceptable automatically gives it commercial value by making those products more usable and therefore more saleable. But in addition, an aesthetic approach does very often reveal solutions to problems which remain hidden to the more material approaches and the few typical examples given of this should establish the value of a visual mode of thought. Visual thinking leads to clarification of forms and to their organization into integrated patterns. It not only does this in the visual sense, but because it is elegant thinking it usually coincides with thinking directed to the best mechanical or structural approaches to a problem. This may be hampered by undue preoccupation with purely technical or material considerations, being as it were too close to the wood to see the trees, which gives the less hampered though equally serious aesthetic approach an advantage. This is not merely fortuitous; it is because form and form-relationships have as much validity and are entitled to the same consideration as is given to questions of what forms are made of or of how they are maintained in relationship to each other. Visual thinking, or aesthetic awareness, is usually quite complementary to engineering thinking and it often leads to the introduction of a rationale and to economies in material and in effort of manufacture of which production and value engineers would be justly proud. Pounds and shillings saved in works costs represent a larger sum in selling cost, so visual thinking can produce not only dividends by making products

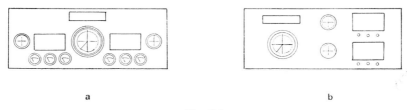

a   b

Fig. 2.1

more attractive and convenient but it can also make them more competitive.

The examples given of the results of visual thinking are typical of innumerable instances of how the addition of an aesthetic approach has been responsible for visually cleaner, functionally better and materially more economical results. These examples are unlikely to be of immediate practical value to the reader in that the circumstances peculiar to each are unlikely to be repeated exactly. As the intention was only to indicate the nature of visual thinking, it was felt better to give actual examples of it rather than to invent theoretical examples.

Example One. (Fig 2.1.) The control panel of a heated-die press was being made visually more cluttered by six electrical meters indicating the loading of the die heating elements. Each meter cost £3–£4 and its scales were standard, fully graduated ones. Enquiry revealed that the

18 THE AESTHETICS OF ENGINEERING DESIGN

press operator did not need to know the exact value of the heater loadings but only to know when they were running above a certain limit, or if there was a cut-out, a circumstance easily indicated by a signal lamp. This led to six expensive and visually-fussy meters being replaced by six small and inexpensive signal lamps with their associated relays and to a clearer and cleaner information area.

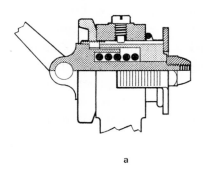

a

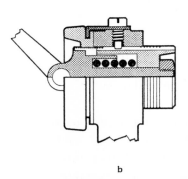

b

Fig. 2.2

Example Two. (Fig. 2.2a.) An adjustment collar was retained against its datum face by means of a rather unsightly single-turn spring and an end washer, a projection of which was bent parallel to the boss and provided with marks which were intended to be read against a datum which was the right-hand face of the boss. This arrangement was unsatisfactory in that visual alignment had to be made across a considerable gap. A re-design of the collar, Fig. 2.2b, led to its being retained by means of a simple tag, with a reduction in overall length, and to the adjustment

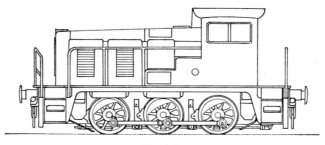

Fig. 2.3

marks being on the part subject to adjustment, where they could be read directly against the reference datum face.

Example Three. (Fig. 2.3.) With the re-design of an industrial shunting locomotive the principal aesthetic aim was to give some impression of length to an object without much length, a dimension consciously kept to the minimum through recognition of the restricted track layouts and minimum radius curves common to the normal working environment of such a product, resulting in an overall proportion of approximately $2\frac{1}{2} : 1$. With locomotives of any kind some visual emphasis and sense

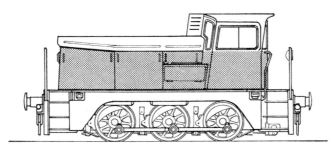

Fig. 2.4

of direction is quite appropriate and a basic aesthetic aim was to make the balance of the value of the basic visual elements strongly in favour of the horizontal, Fig. 2.4. To achieve this, all features tending to disrupt or to weaken horizontality were eliminated or reduced in visual value so far as was possible. A paint scheme was introduced which contributed to an impression of increased length and, associated with this, the cill line of the cab fenestration was slightly lowered and the improved visibility resulting largely from this was an important ergonomic contribution which added to the commercial value of the locomotive. The original vertical radiator cover, a somewhat complex fabrication, was eliminated and an area of slotted metal was introduced which had a stronger horizontal character and which incorporated the two forward side covers, one on either side of the bonnet. This reduced the value of the obvious joints between them and the radiator cover and eliminated the vertical elements formed by the pattern of the louvres punched in these covers. The disruption of the horizontality of the bonnet due to the vertical elements of the notched-out recess for two engine air-cleaners was eliminated and an opening for a single, larger air-cleaner was located centrally in the top surface of the bonnet where it was not seen. The sand boxes with their associated vertical elements were visually incorporated into the side splasher, so adding to horizontality at foot-plate level. The rear-quarter walls of the cab were faired off at a slight angle, adding to the width of the staircase used for across-train movements by the shunter as well as by the locomotive crew, making a contribution, though a small one, to an in-line, dynamic form. Arising out of these various aesthetic aims were the solutions to two problems to which the engineering designers had not found answers. The first was to do with corrosion taking place in the capillary joint between the superstructure and the underbody to which it was tack-welded; the second was to do with the transmission to the driving cab of noise from the engine and the hydraulic converter. The introduction of a positive horizontal break at footplate level as part of the aesthetic aim led to the superstructure being mounted upon a broad strip of corrosion-eliminating, sound-absorbent material, two improvements with real commercial value arising out of purely aesthetic aims.

Example Four. A range of standard machines, lathes, was designed to suit batch-production and while output was at a level of 100 machines per week it was satisfactory, but when it became desirable to increase output to 200 machines per week it became necessary to assemble them in an entirely different manner for which the existing design approach was not suitable. Assembly had to change to a flowline basis which, while it would permit the maintenance of a very high, and guaranteed, standard of accuracy, would not permit the degree of individual fitting and alignment implicit with the existing aesthetic approach, Fig. 2.5. A new approach was necessary and this is an example of where commercial hopes and the safeguarding of a very considerable financial investment were dependent as much upon correct aesthetic decision as upon proper engineering decisions.

The flowline was to consist of two 300-foot lengths of hydrostatic slideways upon which were to float continuous lines of cast iron rafts each 10 feet in length and weighing approximately 3 tons. The planar accuracy of assembly surface essential to maintain the guaranteed accuracy of the machines was 0·0005 in. over 10 feet, a standard which had to be maintained over the 300-foot length of each leg of the flowline. In a normal working week of 40 hours the maximum assembly time for

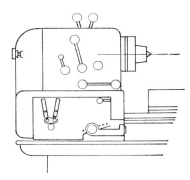

Fig. 2.5

each machine would be 24 minutes only, so the flowline and what was assembled on it demanded the most exacting engineering design. The aesthetic problem was principally one of form-relationships and of selecting an arrangement which would be sufficiently flexible to apply to the variety of sizes of machines in the range and which would produce a reassuring appearance of strength and accuracy, while accepting inevitable size discrepancies between major cast parts without loss of aesthetic quality. The inevitability of such discrepancies must be accepted even where the best standards of foundry practice apply. As this is a known fact, engineering design acknowledging and accommodating it is good engineering design, while engineering design done with the knowledge of this fact but necessitating an unacceptable amount of corrective handwork to achieve an acceptable appearance is bad engineering design. Reference to Fig. 2.6 will show the three principal sources of discrepancy, which are shrinkage, planar distortion and porosity. While good foundry practice keeps these to the minimum it cannot eliminate them entirely and it is not commercially practicable to anticipate the worst situation by specifying a finish-machined size to

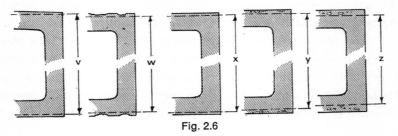

Fig. 2.6

suit the maximum expected combination of them. To do so would be to multiply the number of cuts on each machined surface of the majority of parts to accommodate a minority. It makes better sense to have a design permitting some latitude in the dimension between surfaces which keeps machining and assembly time to the economic minimum. Careful enquiry of the various production interests established which

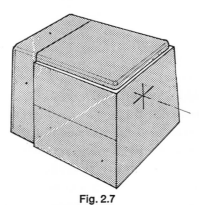

Fig. 2.7

dimensions could be held and which could not and the aesthetic solution was generated from this information, Fig. 2.7. It consisted of two interlocking block forms for the principal visual area at the headstock end and relied elsewhere upon alignments which could be ensured without much trouble, yet which would accept a measure of misalignment without loss of visual validity.

The alignments which could be held were those of the top of the headstock cover and of the end guard, the front face of the headstock and the gearbox cover. The headstock cover was a closely controllable aluminium casting, later a fibreglass moulding, and the position of the end guard fastenings could always be adjusted to suit alignment between guard and cover. The alignment of the headstock face and the gearbox cover depended only upon correct machining of the mating faces and was controllable. The lower edge of the gearbox cover, forming the lower front edge of the headstock block, and the lower edge of the end guard were shaped to produce a continuation of the tray line to integrate the upper, working part of the machine and to define it separately from the lower, supporting structure, Fig. 2.8. This continuation of the tray

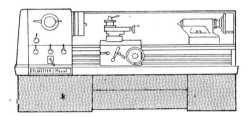

Fig 2.8

line introduced a very strong unifying element and the deliberate breaking of this line, by the two joint lines of the gearbox cover and a coloured area behind the maker's identification, permitted any likely variation of headstock height to be accepted without noticeable misalignment. This alignment, although a strong one and contributing much to the unity of the object, comes well below eye level. Viewed at close quarters it is seen obliquely and is tolerant of some inexactitude; viewed from a distance sufficient to make it read as a principal horizontal line the scale of any misalignment is small and for all practical purposes invisible. This aesthetic solution has proved itself under actual production conditions and is being progressively extended to the whole of the range of machines.

These examples are but typical of many where cleaner and simpler forms and economies in manufacture have resulted in aesthetic qualities generated from objective enquiry into the purpose of the product and the circumstances surrounding its manufacture. No doubt one could show an equal number of simplifications and economies due to value-analysis and a link between aesthetics and value-analysis can be discerned. The difference between them is that the value-analysis result is not always aesthetically acceptable owing to its having been arrived at by ruthless simplification without regard for aesthetic quality. Value-engineering teams and design drawing offices seem often to work in competition with each other, encouraging a climate in which a straight result from one or the other is likely to be accepted whereas the best overall answer is often a synthesis of the efforts of both. It is interesting to recall Professor A. N. Whitehead's definition of style as 'The achievement of a foreseen end, simply and without waste', which is precisely the aim of value-engineers, and confirmation of what a fundamental part of engineering aesthetics is. It is far more important to generate an awareness of this and to encourage a capacity for basic aesthetic rightness among engineers, than to develop purely emotive or stylistic expertise and facility in specialists. The considerable, usually obvious, improvement in graphic quality which applied art can contribute to engineering cannot be denied but it is, unfortunately, usually only part of the cosmetic quality and not of the anatomical quality. Unless that quality can be assured and renewed as technology ushers in new generations of engineering products, the specialists can be no more effectual than cosmetic surgeons doing their best upon unsound and outmoded bodies. Training more surgeons is not the answer, which lies more surely in the production of young, healthy bodies and this, within the context of engineering design, is the engineer's opportunity.

rationalization and optimization

In the comparison of prodigally uneconomic ranges of products with economic ranges of products there is a clear analogy with fussy, unnecessarily complex forms and simple, elegant ones. An unnecessary proliferation of parts usually results from working on a job-to-job basis instead of on a more continuous and better organized one. Variety has virtue and is often justifiable, but where ranges of objects embrace very small and possibly meaningless differences between individual items and where for all practical purposes they might all have been the same size, or reduced to a range of different sizes with a smaller number of steps, it is clear that their existence as a well-related group of objects has not been considered. Unnecessary variety can usually be eliminated without loss of aesthetic, ergonomic, functional or material merit by correctly assessing suitable and justifiable step-values. Such assessment should, of course, reflect the normal steps in discriminatory ability whether anatomical or optical. While selection of those steps is largely a matter of common-sense enquiry and experiment, specialist ergonomic talent is available to be drawn upon where necessary. A great deal of unnecessary effort and expense can often be eliminated by reviewing and rationalizing standard ranges. A further benefit of such rationalization is that with the reduction of the number of size steps the volume of each new step is increased, introducing the possibility of changing from a small-batch method of production to a tooled-up quantity method. Usually there are also further benefits, simpler stockholding and handling, the reduction of the variety of ancillary parts, and fewer finishing and fixing operations. Fig. 2.9 gives the graphic expression of an actual example of an unnecessarily varied range of components, and of the same range after rationalization. The aesthetic quality of a graph line is usually a reliable indication of the material state of affairs represented by the graph; a clean, simple and positive line or curve reflecting a wholesome situation, a jumpy, complex and uncertain line or curve reflecting the opposite. This is another way in which aesthetics can be seen to have a very real connection with aspects of human activity other than purely artistic ones, indicating how aesthetics can have significant commercial value.

The benefits usually arising from an objective and critical appraisal of ranges of standard parts can also be realized by a similar assessment of many complete products. The advantages of having to make, handle and assemble simple rather than complex parts are clear, while the collective effect on a product of many small achievements of simplification and rationalization produces some clear visual improvement. But a great many engineering products remain in a state in which they are, to an expert eye, still crying out for attention. This is generally where the collective effect of a number of minor improvements has been reduced by aesthetic inconsistency, but even where there is consistency the result may often appear as a welcome prelude to a final simplification

26   THE AESTHETICS OF ENGINEERING DESIGN

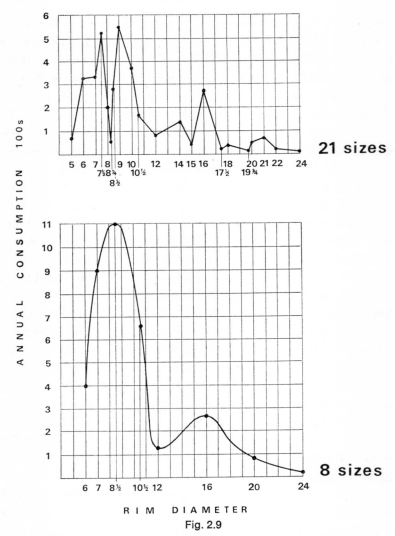

Fig. 2.9

and rationalization that has never in fact taken place. This reveals the absence of a sure, strong aesthetic purpose, one that accepts the minor and more easily recognized benefits of simplification as natural preliminaries to an obviously better final state of affairs. Tenacity of purpose here is therefore just as important as understanding of purpose; the real goal is so often in sight, yet apparently not seen, and the culminating steps that might have made a great deal of hard groundwork flourish are not taken. Of course there must be some products where for convenience of making, or in assembly, complete sub-assemblies which can be tested must be retained. Here the final whole object may not necessarily be reduced to the simplest and most economic structure, but it can usually be visually consistent and cohesive, provided that some concern is shown for those qualities.

But even where strong aesthetic ability is available, visions of ultimate simplicity, compactness and unity must always be tempered by recognition of just how far it is desirable or economically possible to go. Some sub-assemblies containing numerous shafts and gearing, or mechanical parts with critical relationships, need to be tested before being released by their assembly sections. Common-sense and economy often lead to the structure of such assemblies being combined with their outer covering or housing; as it were, the skeleton with the flesh and the skin. But even where secondary sub-assemblies have to be accepted as separate, visible parts of the whole, they need not be any less well co-ordinated than are the torso, head and limbs of the human body. Where their structure is not also their overcoat they might be assembled beneath a common, unifying cover and where this does not add to overall cost and where a more complete simplification is not possible, this is often the best solution. There are, however, many instances of where products remain as, apparently, an haphazard collection of sub-assemblies lacking in unity more through convention than out of any practical necessity. Such passive resistance to change and reluctance to try any line but the one of least resistance needs to be discouraged, particularly where advantages other than improvement in aesthetic quality are possible. Compaction, reduction of required working-space, easier forms to make, finish, assemble and maintain, these are all solid, sensible commercial benefits possible where closed minds, convention and prejudice do not preclude them.

The example shown, Fig. 2.10, is typical of objects susceptible to this kind of visual reorganization and it also illustrates how the problem of essential, though often visually incompatible, components can sometimes be dealt with. Such items are often produced as self-contained units, but when incorporated into a system with other items their individual packages may become unnecessary, indeed they could prevent proper integration. By discarding the package, an associated component can become an integrated component with considerable visual improvement and economy in overall space required. Fig. 2.10a shows a shoe-forming machine which was originally bench-mounted with, standing at the rear, a refrigerator unit for cooling one of the working heads (not shown) which project from the front face. A simple re-assessment of this machine, Fig. 2.10b, produced a unified object with the refrigerator unit, without its own cover, contained in the housing beneath the heads. The electrical controls, formerly in a box on top, are also integrated. The achievement of a maximum result, where that is permissible, is most likely if the simple basic principles of perception are adhered to,

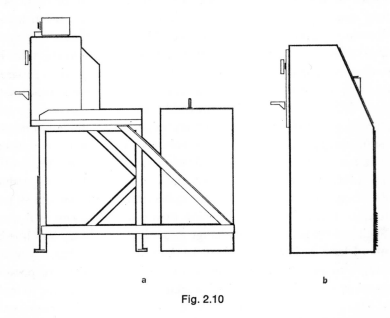

a   b

Fig. 2.10

if you just ask 'What is the simplest form consistent with convenience in making it?', 'If some visual complexity seems likely, what is the virtue of it?', 'Will it in fact stand up to objective analysis and examination?'" and, should there be some virtue in complexity, 'Will it outweigh the virtue of there being no complexity?' This simple technique of enquiry seems to have been one of Leonardo's strengths and it can often result in quite radical improvement out of all proportion to the effort involved. The development of a questing, non-conformist outlook need not constitute an assumption that products are never so well-designed as they might be, but an enquiring, critical approach based upon a lot of Why's? and generating but a few sagacious Why Not's? will usually result in the correction of practices responsible for less-than-maximum product performance or a less-than-optimum suitability for manufacture. All design is a compromise of conflicting requirements and the most satisfying results are those where the priorities of the conflicting needs have been correctly assessed—which is, unfortunately, too seldom the case.

THREE

perception

WHY SHOULD engineers with their many other concerns be interested in perception? They should be interested because they are creators of a great deal of what is perceived, and an appreciation of the mechanics of perception can only be helpful to them.

Take any object and look at it. The eye does not understand anything it sees. It is simply an optical relay station observing visual stimuli and reporting them to the brain, analogous not so much to an ordinary camera, recording latent images upon sensitized film, as to a television camera interpreting a visual scene into a pattern of impulses, or coded signals, which are passed on to the brain via the optic nerve. The brain, upon receipt of this information, compares the stimuli with its stored experience of similar stimuli patterns and tries to identify the pattern the eye is seeing at that moment so that it becomes a pattern of significance. It does this in a millisecond and it may be able to make identification completely and without hesitation, but it may find the pattern similar to several stored patterns and be able to concede identification only with reluctance, after recognition of the context within which the object observed is associated and with other visual clues. So already there are two factors of concern to engineers. First, the realization that confident recognition, acceptance and full use is dependent upon clarity of perception, which is itself dependent upon the firmness of the visual statement. Second, the existence of an irrational factor in perception leading to overtones of doubt in identification which can linger on, to the disadvantage of what is being perceived. This is a common experience; we may be looking at a metal bracket, Fig. 3.1, which may be perfectly strong but, because it has reminded us of some plant or vegetable form which we know to bend or to yield easily, we cannot be at ease with the observed object within the context of its use or applica-

a   b

Fig. 3.1

tion. The physical qualities of objects may be perfectly adequate and there may be no rational reason for our not being happy with them, but because of this irrational factor in perception we are completely satisfied only when the object's visual presence is consistent with purpose, function and material. In short, as much attention must be paid to an object's visual presence as is paid to any other consideration in making it; if it is not, there might be little point in having bothered to make it. Unsatisfactory presentation of good thinking is like having good lines spoken by a poor actor, the alternatives being to have the lines spoken by a good actor, within the context of our subject an aesthetics specialist, or for the author, the engineer, to speak them himself. This would give him proper control over inflection and emphasis and would remove any misinterpretation of the overall intent, but it does require the engineer to have the necessary ability in visual presentation.

The retinal images of anything we observe are inverted, like the image on the focusing screen of a camera, and the reason why life does not appear to be all upside-down is that the brain makes the necessary correction. It also has to do a great deal of other work in perception, the most important task being that already referred to, the identification of what is being observed. The avoidance of visual arrangements likely to evoke inappropriate mental images is a natural responsibility of an engineering designer, a sufficient reason why he should have some awareness of a subject which at first may not seem to have any very strong connection with engineering. It is helpful to know about, though not necessarily fully, the Alpha rhythms. These are rhythmic oscillations at a frequency of about 10 per second which transform spatial patterns into temporal patterns, into a succession of signals in time. This transformation of spatial into temporal co-ordinates is roughly equal to scanning and where the sensory data is clearly defined and the signals to the eye are positive, the result of scanning is rapid and precise. Where the sensory signals are indefinite, scanning is sluggish and confused, and this is the reason for the emphasis throughout this book upon clear, strongly-established visual form. In perception, that fundamental human ability, we are all obliged to be party to confidence trickery and we are all, whether we like it or not, operators in a gigantic con game. In the pattern read as a white C on a black ground, Fig. 3.2, the eye sees only approximately 15 per cent black and 85 per cent enclosed white space; the brain interprets the pattern as an overall black area upon which is superimposed a large white capital C. It should, however,

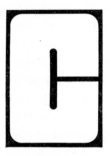

Fig. 3.2

be appreciated that the use of the word 'trickery' here implies only the conscious representation of fact without necessarily any wrong intent. But any interpretation of fact is open to misrepresentation, because in order to recognize and to identify positively what we see we have to compare the visual stimuli with similar sets of stored stimuli in the experience-memory. This memory store is usually sufficiently adequate to enable us to be sure of the class of an object, though not necessarily sure of various qualities it may possess and an honest, objective intent is thus likely to be overlaid with hope or fear as well as leading to the same observed object being interpreted differently by a number of people.

To this ever-present liability to deceive ourselves can be added the fact that designers set out to organize deliberately what is seen to produce a certain overall interpretation, so in addition to being the potential victims of a natural phenomenon we are also the potential prey to a designer's will. Of course, there is nothing undesirable about this when the visual emphasis of good qualities leads to actual experience of them, which might not be the case were the object characterized by visual evidence of bad qualities. It could be undesirable however where, to satisfy some purely material interest, or human conceit, something fundamentally bad was made to be interpreted as something fundamentally good. A designer is thus for good or bad a con man and, as with all human behaviour, there are certain tolerances of acceptability working within which is considered acceptable, while straying outside them is considered anti-social and unacceptable. The establishment of these tolerances is based more upon arbitrary psychological and material grounds than upon strictly ethical ones, and in commerce cynical promotion is permitted to deceive by playing upon people's vanities and credibilities and their readiness to conform to established social patterns. How far a designer is prepared to go in the organization of visual stimuli to encourage certain interpretations of it is a matter for his conscience alone, but whether he stays within the accepted tolerances or oversteps them he cannot escape being a manipulator of confidence—in the best meaning of the phrase, a confidence trickster.

An awareness of this truth can assist in the recognition of, and in the deflation of, imposed self-opinion or deliberate attempts to mislead, as well as, of course, of any honest endeavour merely to present something to best advantage without any deceptive over-emphasis.

irrationality

There is some emphasis in these notes about perception upon the inescapable element of irrationality in it, and which can distort or actively undermine a great deal of effort in the realization of hardware. This can operate in two ways. The first is in identification, when, if the observed form strongly resembles other forms, certain known qualities of those forms may be attached to the observed form, however inappropriate they may be. The second is apparent structural improbability, when certain physical defects may be read into a form which is materially quite satisfactory. Examples of these aberrations are, when we feel certain that a form will readily bend, wilt or sag simply because it reminds us of forms known to do that, in spite of the observed form being made of materials that would not permit it. Also when we feel positive that certain parts of a form are about to slide off or to break out of other parts although, again, the material and the construction would not permit this to happen.

It is of no use thinking that this quite irrational kind of thing cannot and will not happen; it can and will. It is equally of no use to think that we should concentrate upon making human beings more rational and remove the necessity of having to bother with little tricks of visual arrangement merely to prevent them from arriving at wrong conclusions about what they are seeing. Getting human beings 'straightened-out' and removing problems arising out of perception is a charming thought, but when they cannot even be persuaded to cease doing all kinds of terrible things to each other there seems to be little hope of re-educating them in the complex and sensitive activity of perception. We are almost certainly better advised to accept the lesser problem, which consists of accepting human beings as they are and arranging what we do to suit them. That poor aesthetics can be as damaging as using poor materials and workmanship is a fact that is not always realized, but if the aesthetic quality of a product does not support its performance potential it is equivalent to employing unsuitable materials or insufficiently precise methods of manufacture. Where highly accurate relationship is essential, but where the form does not suggest it, faith in performance must suffer and a great deal of effort, and investment, may be placed at hazard. It is quite possible to take a form that is materially, structurally and functionally adequate, an adequacy reflected by its visual characteristics which could evoke a feeling of complete assurance. Then, without prejudice to those qualities, it is quite possible to alter the visual characteristics of the form so that it is no longer reassuring, perhaps even downright disturbing. This transformation, this liquidation of confidence, could be the result of engineers themselves making quite satisfactory engineering changes but without attaching any significance to the effects of their activities upon the form's visual characteristics. The net result could be 'as you were', with any improvement in mechanical performance cancelled out by a loss of confidence in the product.

Of course, the reverse could be the case and a visually unconvincing form could be made more convincing and acceptable by suitable adjustment of the visual characteristics, and this is where the work of engineering designers can be seen to be potentially as significant as the most important technical specialists, whose work they can either reinforce or undermine. Any serious engineering project evoking a wrong

PERCEPTION 35

association of ideas and suggesting adverse, even derogatory, qualities is a pathetic thing and could be a dissipation of considerable effort and investment. It could happen that a machine or a piece of equipment has been made with adequate structural strength which is visibly apparent when the machine can be seen completely, Fig. 3.3a. But if, in order to bring the centre of work-operation down to operator level or to make the work-handling height to suit existing ancillary equipment, it becomes necessary to sink the machine into the floor, Fig. 3.3b, what remains visible may no longer appear to satisfy structural requirements. Where work accuracy depends upon this, as it often does, this could be a serious matter. First impressions, though overall and superficial, can

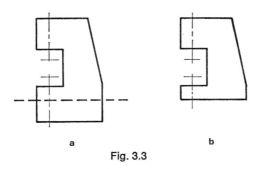

Fig. 3.3

be critical and decisive and in the case of this hypothetical example some corrective treatment would be justifiable and perhaps commercially necessary. To achieve such correction without interfering with function or adding cost is not always easy, but any reasonable addition to manufacturing cost must be properly weighed against any prejudice to adoption were the correction not made. Of course the most satisfactory thing is to avoid such situations, but this can only be done when engineering designers can foresee the visual implications of their work and this need only call for an overall, but sound, appreciation of aesthetic cause and effect and not necessarily any great executive expertise.

Experience of assessing the aesthetic and ergonomic qualities of design projects of fourth-year students at a senior college of technology shows that the best results emerge where students have stuck to sound, basic engineering principles and qualities and to good common-sense. The worst results emerge from conscious effort to achieve aesthetic quality by employing idioms that are not properly understood, like words used out of context when the grammar is not understood. It is clearly much better for engineering designers to get the basic principles right than to try to achieve any great facility in the execution of aesthetic detail. That can always be attended to by specialists if need be, but what the specialist is able to do must always be controlled by the basic material upon which he has to work. Whether the future of engineering is going to be one of the multiple specializations or of completely competent, self-sufficient engineers does not alter the fact that engineering designers should be capable of a reasonable standard of

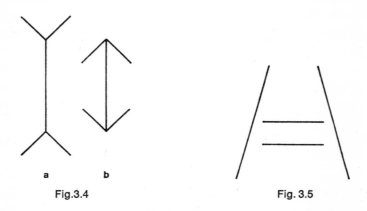

a   b
Fig.3.4   Fig. 3.5

aesthetic quality. An ability to ensure the fuller meaning of aesthetics is of far greater importance than any facility for achieving superficial, emotive, visual effects, for in addition to contributing artistic and ethical 'rightness', aesthetic quality can have very real commercial significance in the avoidance of unfortunate perceptual effects and in the underwriting of the total effort and expenditure behind any engineering enterprise. Reference to the illustrations showing two well-known illusory optical effects, Figs. 3.4 and 3.5, should remove any doubt that what the brain makes of something does not necessarily agree with what the eye sees.

The vertical lines in Fig. 3.4 are of the same length although the line in Fig. 3.4a appears to be longer than that in Fig. 3.4b. A physiological reason for this is the extension of visual interest by the outward pointing chevron lines at top and bottom, whereas in Fig. 3.4b these lines are inward pointing and tend to reduce the spread of visual interest. A psychological reason is that Fig. 3.4a is reminiscent of the corner of a room, a close environment in which dimensions are not reduced by distance, whereas Fig. 3.4b is reminiscent of the corner of a building, something larger in scale than the room and which is reduced in scale by distance.

With the Ponzo illusion, Fig. 3.5, the oblique lines suggest a plane in perspective and the uppermost horizontal line being apparently farther away from the observer is perceived as being longer than the lower horizontal line.

It is owing to our being obliged to resort to analogy through having to compare visual stimuli with stored experience in order to be able to identify anything, that we are apt to try to make the facts fit the theory,

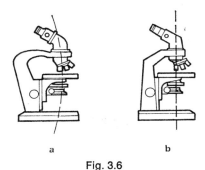

Fig. 3.6

rather than the other way round. Manifestations of this element of irrationality leading to the attachment of wrong qualities are common and the example shown in Fig. 3.6 is typical. It shows an original proposal for the limb and base for a polarizing microscope intended for petrological and metallurgical work where the degree of accuracy has to be even higher than it does for pathological work. The wilting form of the limb and the lack of visual unity between the base and the cylindrical stack comprising the workstage, analyser and optical head, Fig. 3.6a, failed to suggest a highly accurate relationship, although it would, of course, have been present. Stiffening of the limb, Fig. 3.6b, and the unification of the shapes of the limb, base and worktable restored a visual impression of accurate relationship, making the visual presence of the product properly support all the tremendous instrument-making skill that had gone into it. Examples of imagined structural inadequacy or improbability are also common, and occur with all things appearing to be top-heavy, likely to bend, break or fall off. They occur also in less obvious but equally devastating ways, as with overhanging or cantilevered forms, Fig. 3.7, where there must be some physical or optical break between the supporting structure and the projection. A clean vertical break will give the impression that the projecting part is likely to snap off, or to slide down the face of the supporting structure, Fig. 3.7a, unless the projecting part is visually made part of an inner supporting structure by structural or optical links, Fig. 3.7b, or unless it is made to appear to be somehow physically interlocked or connected.

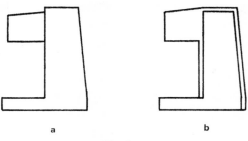

Fig. 3.7

An adequately strong metal structure may appear to be otherwise if it happens to remind us strongly of the form of something we know to be otherwise. Where considerable sums of money and effort have gone into ensuring unusual degrees of rigidity, accuracy or some other special virtue, it is not sensible if this achievement is partly or wholly neutralized by forms evoking quite opposite and adverse qualities. Any emphasis upon the irrational element of perception is due to its being the root of most that is intangible in aesthetics and as this book is directed principally to engineers and to engineering designers accustomed to tangible, absolute values it is important to draw attention to it. Because the permutations of sensory data from which perceptions are built are infinite, there can be no clear-cut rules in aesthetics, there can only be some general principles with the resolution of them dependent more upon feeling than upon precise, calculable understanding. This is by no means to suggest that aesthetics is a form of anarchy, without order and where every opinion is as good as another; rather does it indicate the power of feeling which, conventionally, does not rate very high among engineering values. Custom and habit have given rise to the assumption that what cannot be easily classified, apprehended in quantitative terms and interpreted in conventional imagery, is of no particular significance.

basic visual elements

No consideration of aesthetics and perceptual effect can avoid what is known in artistic terminology as the figure and the ground, interpretations of which will be found in the glossary of terms on page 121.

Variation of the figure on the ground will vary the perceptual effect and where the figure is made up of a number of visual elements, as most engineering products are, the overall perceptual effect is the outcome of the competition for your attention by those elements. Wherever there is inequality of effect there is competition, hence the possibility of conscious organization to achieve a desired effect. Aesthetic quality depends largely upon such organization although it cannot be determined by it entirely because the quality and elegance of the concept, the thinking upon which any visual display is based, must have its share. One real difficulty with aesthetics is that it has no language of its own and we are obliged to fall back upon analogy with other human experience, either material, physical or sociological. Because aesthetics is concerned with sense-reactions, with sensations, most analogies tend to be with familiar experiences. We describe pain as shooting, burning, throbbing, irritating, nagging and so on and similar analogies are made to describe aesthetic effect and manoeuvre, where we say that the arrangement of the visual elements of a form can give it direction, movement, balance, weight, robustness, delicacy or some other experience-based quality.

Visual elements are, of course, abstractions without weight or movement, although they may be the principal visual characteristics of hardware having those qualities, while the weight and movement ascribed to visual elements need not always represent what could occur with the hardware. This may sound somewhat confusing but this is the nature of aesthetics and the folly of attempting to be dogmatic about aesthetic matters is suggested by the consideration of a C-frame struc-

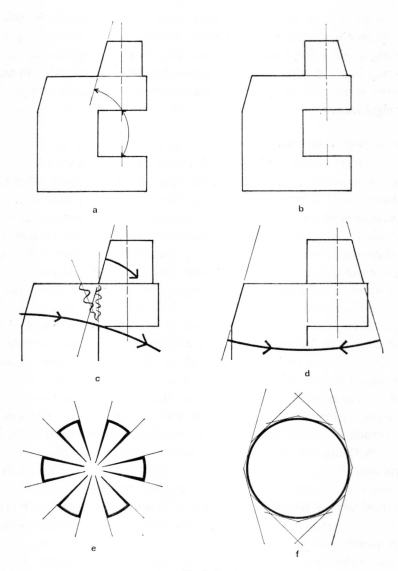

Fig. 3.8

ture, Fig. 3.8. This would have an open throat to permit the admission of work between the upper and lower limbs, between which pressure is generated by the work process. The structure might be crowned by a protective cover, Fig. 3.8a, over hydraulic cylinders and piping and to relieve the bulk of this and to reduce its visual weight some tapering may be considered desirable. It might be thought that by sloping the rear face of the top cover a line would be introduced normal to, and opposing, the imagined line of stress arising from the work process; considering only the dynamics of the structure this could be an acceptable intention. But if the static structure only was considered it might be better to slope the front face of the cover, Fig. 3.8b. These differences in approach arise from the quite irrational effect of the oblique line. When this is at the rear of the cover it certainly opposes the thrust of the pressure generated in the throat, but, because it appears to be hinged almost at the upper side corner of the throat, it could induce a suggestion of structural weakness, Fig. 3.8c. In general, oblique lines running in a similar direction, with no like lines to oppose them and gravity, apparently, assisting them, introduce an unstable, disruptive effect, Fig. 3.8e. On the other hand, contra-sloping lines, Fig. 3.8d, have a compressive, stabilizing effect and are associated with structural solidity and with complete, easy-to-identify forms, Fig. 3.8f. Of course, the colour scheme, the form of the work-heads within the throat and other visual details should determine what it is best to do in these particular circumstances, but a safe decision might be to locate the sloping face at the front of the top cover.

Further references to figure and ground will be found in the part of this book dealing with surface display and its arrangement, but attention should be directed first to the broader matter of round form and to some consideration of the basic linear elements by which it is characterized.

Form, in the aesthetic context, means shape or arrangement of parts. It generally implies a three-dimensional state but an appreciation of form is best begun with an appreciation of shape, which can be two-dimensional and this makes desirable some understanding of the visual elements from which shapes are constructed. These elements are the basic linear ones, the horizontal, vertical, oblique and curved lines, Fig. 3.9. The interpretation of a line in the aesthetic context is an extension of length without breadth, an abstraction, and it may at first be difficult to see how abstractions can find any place in three-dimensional form. Lines can be perceived as the contours or boundaries of forms, as definite changes of plane and as dominant visual axes, those lines 'felt'

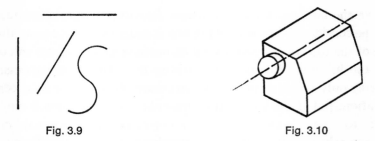

Fig. 3.9 Fig. 3.10

but not seen, Fig. 3.10. The four basic elements have no fixed relative values because those can alter according to the way in which the elements are used and to the implications their use evokes. In Fig. 3.11, two arrangements of the same three elements, two oblique and one horizontal, evoke antithetic associations. The first arrangement, Fig. 3.11a, suggests a pyramid and, as we know that the Egyptian Pyramids are enormously solid and sit heavily upon the earth, is essentially a

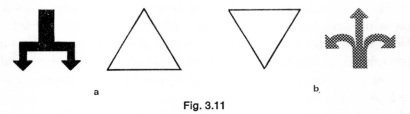

a b

Fig. 3.11

static one. The second arrangement, Fig. 3.11b, suggests a free flying kite or an unstable form likely to overbalance at any moment, a situation charged with potential movement and essentially a dynamic one. Although this may make the situation seem to be too indefinite and not conducive to orderly arrangement, some rough relative values are sufficiently constant to provide a basis for objective organization.

The horizontal element, Fig. 3.12a, is physiologically somewhat easier to perceive than the others and tends to be the norm, encouraged by

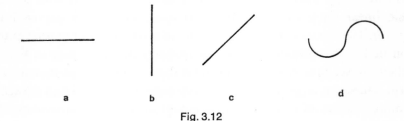

a b c d

Fig. 3.12

the facts that most human activity takes place within a fairly clearly defined horizontal stratum and that it is easier to scan horizontally than in any other direction. The sense-impression of horizontal elements is of nothing unusual happening and of the least demand being made upon the perceptive mechanism. The horizontal element has thus tended to become an analogue for quietness, passivity and peace and in symbology both ancient and modern it is used as the symbol for the female element, for subtraction and for the negative potential. Leaving aside the matter of thus characterizing the reproductive member of the species, these associations are usually with quiet, passive and static things.

The vertical element, Fig. 3.12b, is a change from the norm: visually it can be a welcome relief to the monotony of the horizontal and it makes a rather stronger impression. Somewhat more effort is demanded in scanning it and it registers as something a little more important and, if its scale is sufficient, it can prise the eye out of its normal horizontal stratum thus attaching more importance to itself. Even when its scale is small, say a tree, mast or spire upon the horizon, it can still register in our consciousness in spite of the apparently overwhelming competition from horizontal elements. Symbologically it is used for the more active male element and it converts the passive horizontal element into the symbols for addition and for the positive potential; all more dynamic and positive things.

The oblique element, Fig. 3.12c, registers as something more important still as it requires a little more effort to perceive, muscular activity now involving two sets of ordinates; there seems to be something more important happening and the sense-impression is that much stronger. Perceptually it is the most dynamic of the elements so far and symbologically it was associated with things most important to human life in the days of sign language, such as the sun's rays, lightning and many important cross signs.

The curved element, Fig. 3.12d, that is, any non-straight element, makes the maximum demand upon the perceptive mechanism. As with the oblique element two sets of ordinates are involved and their ratio is continually changing though not necessarily at a constant rate and our attention is not only arrested by the quasi-oblique but there is a little more work to be done in perception. Symbologically we find the curved element associated with the most important things, such as God, creation, unity, world-without-end and so on, while the changing quality of the development of three-dimensional multi-curved surfaces offers greater visual interest than do the simpler, more readily-perceived developments of a flat surface.

There are four basic relationships, Fig. 3.13, of shapes and forms constructed from the basic linear elements; they are through size, shape, location and surface characteristic, that is, colour and/or texture, any relationship being, of course, through similarity in any of these things. What is said elsewhere in this book about visual unity is relative.

visual competition

The relative values of the basic linear elements and the competition implicit with difference admit the possibility of conscious organiza-

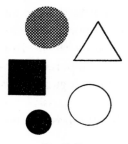

Fig. 3.13

tion to produce desired visual effects, in other words, of conditioning perception. There is nothing arbitrary about this, it can be quite objective and its value in engineering design is that it permits a considered approach to be made to some problems to which the only tenable solution may be the aesthetic one, but where, because of the popular misconception of aesthetics as being esoteric and not positive in the engineering sense, an aesthetic approach is not made.

Visual competition can arise from the characteristics of the basic linear elements and from forms composed of them, Fig. 3.14a, and from the positioning or isolation of forms in such a way as to influence visual attention, Fig. 3.14b. It can arise also from the scale or the boldness of forms and here strength of line, visual weight and surface characteristics can be as important as physical size, Fig. 3.14c. Indeed, in visual matters

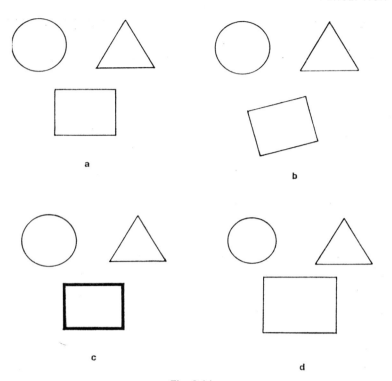

Fig. 3.14

generally, quality is invariably superior to quantity in influencing effect. Size, Fig. 3.14d, can however come into the picture and visual competition can sometimes extend to a complete reversal of the relative importance of figure and ground, Fig. 3.15, i.e. to one form becoming more important than another form upon, or against, which it appears. This can be important in engineering where a secondary form should not be permitted to dominate a primary form for functional or structural reasons, or where what is visually a minor form, although function-

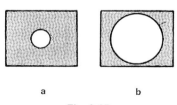

Fig. 3.15

ally of major importance, requires to be made predominant. In aesthetic terminology this reversal of the roles is where a superior ground and an inferior figure become a superior figure and an inferior ground.

It is obviously important that any desired visual priority is known and that all design decisions are calculated to maintain that priority, and this is why mechanical designers whose work may not be visible on the surface must be sufficiently aware of desirable overall aesthetic aims to avoid sabotaging them by decisions made in respect of function and structure alone. To believe that aesthetics can be separated from the normal hour-to-hour working of engineering design to be applied at the end of the day, week or month is quite ludicrous. It would be, as far as ensuring human values is concerned, working from the outside inwards to discover one channel after another blocked and resulting in what should have been a fine, spring tide becoming a miserable, inadequate neap affair. Realization of the truth of all this is of the utmost practical importance if the designer is to know whether to adjust the figure to the ground or the ground to the figure. It is equally important to realize that once basic anatomical design decisions have been taken in this respect, no amount of frantic surface cosmetic adjustment can alter things if alteration becomes clearly necessary. Uncertainty of aesthetic aims can only result in mediocrity of appearance and while the participation of an aesthetic specialist can minimize mediocrity, it cannot eliminate the possibility of it; that is only possible where engineers and engineering designers are aware of the human aspects of their work.

sense-impressions

While sense-impressions and the sense-memory are subjects not to be tackled lightly by someone other than a psychologist or a physiologist since the recognition and remembrance of visual patterns is still something of a mystery, some rough understanding is possible through following Leonardo's principle of always going back to square one. In looking at an object the first impression is perhaps that it is too long, but on looking again it may be decided after closer examination that it is not so long after all. Why then did it seem so certain that it was a long form in the first place? What may have happened is this. While the

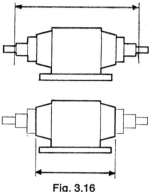

Fig. 3.16

main part of the form of the object may not have been very long, Fig. 3.16, there may have been small features acting as extensions which the eye took in as part of an overall appraisal, and there may have been features enabling the eye to read more surely and easily in one direction than in another. The brain probably formed the impression that the eye was seeing a long form, because scanning was easier in one direction than in another and because it was aware of the journey of the eye out over the extensions. This explanation may sound improbable, but this is what seems to happen. If it sounds irrational that is because it is irrational. Irrationality enters into perception quite strongly, a fact which becomes very obvious in any consideration of attenuated forms or forms with indefinite junctions or divergences. With an attenuated form, Fig. 3.17, the eyes seems to travel along any parallel part of it, recording a constant value which it relays to the brain, until it comes to the tapering, semi-circular or diminishing part of the form when the value of the signal falls while the eye scans on with progressively less to report. Unless the scale is large enough for the form to be read as an

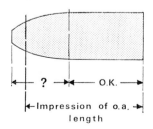

Fig. 3.17

48 THE AESTHETICS OF ENGINEERING DESIGN

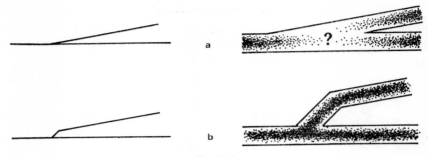

Fig. 3.18

area rather than as a linear figure, the brain seems to balance the reduced signal value against the extension of scan after the commencement of the signal fall-off. The result is that the sense-impression is one of the end of the form being somewhere along the attenuation, certainly beyond the parallel portion of it but not at the actual extreme. What seems to have happened is that uncertainty has been introduced, through there having been no definite termination of what had been a constant experience arising from the uniform width of the observed form. There has in effect been a mental hiatus and for this reason the round ended stroke produced by the rotating cutter of an engraving machine does not produce the most clearly defined letter forms, unless their size is sufficiently small for this effect of uncertainty to become negligible.

With the sense-memory it seems to be certain that unless visual values are clearly and strongly established, the brain will be obliged to make an average decision upon what is reported to it and the resultant sense-impression may not agree with the object itself were it to be measured. The visual uncertainty giving rise to mental hiatus occurs with forms having slowly diverging elements, Fig. 3.18, like a very slowly diverging road junction; the eye is into the divergence before the brain is aware that there is one and it is some time before it seems to be certain of which track is being followed. The result is mental uncertainty, hiatus. This occurs strongly with thin wedge or featheredge forms, Fig. 3.19.

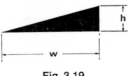

Fig. 3.19

They are best avoided for visual reasons, while materially, too, they are undesirable in that they impose an undue need of attention in their realization owing to one dimension being entirely dependent upon the other, for unless the width is just right the height will be affected.

Another way in which this phenomenon occurs is experienced when reading a circle or a curved line in alignment with straight lines, Fig. 3.20. The eye reports the terminations of the straight lines and the brain

Fig. 3.20

can be reasonably positive about their length, but with the circle or curved line the approach to the datum at the maximum point of curvature becomes progressively less positive, while establishment at the datum is so brief and the retreat from it so progressively marked that the brain seems willing only to concede recognition of something less than the maximum diametric travel. In short, it concludes that the round forms are smaller than they in fact are and, although they may be resting precisely upon a datum, that they appear to be failing to

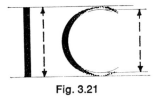

Fig. 3.21

reach the datum, Fig. 3.21. Lettering artists always compensate for this effect by making the overall dimensions of circular or part-circular letters greater than the height of the straight-element letters, and with hardware this effect can result in an impression of a form not having fully arrived, of still sneaking into place while the rest of the company is already present and on parade. Quartic or circular covers, frames or dials aligned with square or rectangular ones (Fig. 3.22a) should always be dimensionally larger (Fig. 3.22b). Just how much must depend upon other factors in the visual equation such as tonal contrast, any three-

50 THE AESTHETICS OF ENGINEERING DESIGN

a   b

Fig. 3.22

dimensional effect of frames or bezels, and so on. This is yet another reminder that the correct answer to any aesthetic problem can only arise from the facts of a situation. Unless the facts can be altered any solution generated must fit the situation. It is a reminder, too, that engineers must have an awareness of the aesthetic quality of their work if the objects they produce are to remain within the limits of aesthetic acceptance.

FOUR

form

CONSIDERATION OF form soon indicates that there are two separate, though strongly related manifestations of it. There is the internal form of the underlying basic structure, the anatomical embodiment of form, and there is the external form of the visible shape, the surface, the cosmetic embodiment of form. Each is as important as the other, but there is a difference of importance in the order in which both are achieved. It is most important that the basic aesthetic, the anatomical quality, should be right before any effort is devoted to the cosmetic quality; for unless the first is right the second can be no more than the icing on a poor cake. In addition to getting this sequence right there must always be consistency in the characteristics of both, for without consistency the result will be a wolf in sheep's clothing, or vice versa, while considerations of composition, proportion, direction and other unique qualities apply equally well to both. The increasing tendency to relinquish the treatment of the external form or the cosmetic quality to others, leaves the engineering designer solely responsible for the internal form, the anatomical quality, and this is a sufficient reason why he must be aware of the visual consequences of his discharge of that responsibility. Indirectly it is the reason for this book, in which factors common to both internal and external form are dealt with together and only factors peculiar to external form are separated and treated as considerations of surface.

composition

The composition or arrangement of the parts of a visual pattern is, with engineering design, naturally very strongly influenced by the requirements of mechanical function and structure. This can be a blessing and a help in that it usually ensures a purposeful, practical and logical effect, but it does not always make the achievement of a purely artistic effect easy. It is necessary to recognize that engineering has its own aesthetic, as does any artefact, and there is thus no reason at all why those who really understand engineering should be any less well placed in the realization of that aesthetic than are artists. Basic functional and structural requirements are therefore the proper foundations for composition in engineering design, not aesthetic theories, which can often salve a situation but should never be permitted to dictate it. Except where the basic requirements are overlaid with purely commercial considerations and so possibly made unbalanced, engineering requirements are generally well-balanced ones and result in acceptable compositions. The qualities required of compositions of basic, anatomical form are obviously ones suggesting material unity and strength and indicating as clearly as possible what the functional purpose is. Unless these qualities are achieved in the anatomy it is usually impossible to achieve them cosmetically, but guidance upon composition cannot really extend much beyond a warning to avoid effects of physical imbalance, weakness, clumsiness or indecision. Here the aesthetically ubiquitous resort to analogy is the surest guide. A beefy ox of a man may suggest great strength and physical reliability but not precision, accuracy or agility; on the other hand an over-refined, precise and meticulous person may not evoke any assurance of a capacity for endurance. The merit of a trim athletic figure carrying no excess weight, with adequate muscles in the proper places, is apparent to all and once this habit of mental analogy is developed it will be found to be applicable to the most unlikely situations and capable of resolving problems of the expression of the most unique blends of qualities. This may not seem to be of much help to anyone hoping for more detailed guidance, but observance of the basic engineering requirements, plus a concern for visual quality, should result in an acceptable engineering aesthetic.

Elsewhere, as in Chapter 2 dealing with the commercial value of aesthetics, there are examples of the recognition of the intractability of most engineering materials and of the inevitable discrepancies implicit with some basic processes. Observance of these facts is essential to the realization of a proper aesthetic for engineering. Design which observes these facts can be good engineering design, while design which is aware of the facts but ignores them is poor engineering design.

It is on the whole easier to supply more closely defined guidance upon graphic layout and the realization of the cosmetic quality in general than it is to offer precise guidance upon the realization of the anatomical quality, which leaves getting the basic form right very much a personal challenge to engineering designers. A challenge not to be underrated, but one made less difficult by the susceptibility of the solution of most aesthetic problems to simple analogy, as outlined in the chapters on perception and aesthetic manoeuvre. An appreciation of the possibility of direct analogy of visual problems with more familiar, better understood phenomena may for many throw a fresh light on aesthetic matters.

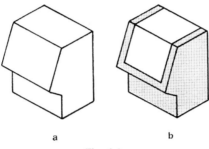

a b

Fig. 4.1

surface

Surface describes the external form, or the visible shape, and is the cosmetic embodiment of form. As suggested above in the introduction to this chapter, the quality of the external form is of no less importance than that of internal form, but this is contingent upon the internal form, or the anatomy, being right. Unless this is so, this aspect of the total form can be quite superficial and as misleading as anything cosmetic can be. A surface is either an essential visible part of a form, Fig. 4.1a, or an area applied to or carried by a form, Fig. 4.1b, and it is important to distinguish what the intention is before making arrangements to realize it. In the first case, priority must go to the ease of reading and identifying the form of which the surface is a part. In the second case, it is essential that the carrying form has its own structural validity, permitting the surface under consideration to be dealt with in its own right so that it cannot be perceived as a false wall or fascia physically or optically weakening the carrying form. Sound realization of a carrying form enables any surface carried on it to be considered in terms of the colour and texture best suited to it and without prejudice to structural validity. Considerations of surface must therefore be either in respect of its implication as a necessary structural part of a form, that is, of its anatomical quality, or in respect of its characteristics of colour or texture, its cosmetic quality. The decision should be dependent upon the functional role of the surface, and complex surface developments that are visually very strong can compete with the form of which the surface is but a part and can, in some cases, amount to secondary forms challenging the primary form.

Where the development of a surface is due only to the accommodation of functions contained within a form and is under the control of the designer, the simpler and more unified it can be the better. But where disruptive development is necessary and due to applied acces-

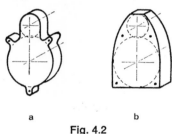

a b
Fig. 4.2

sories, the form of which is not under the control of the designer, the possibility of a unifying cover which would restore the structural validity of the principal form could be considered as the most hopeful solution. What is disrupting the surface may, of course, be in itself aesthetically acceptable and there may seem to be no good reason for concealing it, but it may not subordinate itself sufficiently to avoid competition with the overall visual intent and covering it up could be justified. The surfaces of many engineering products are often unduly visually complex as a result of the designer having followed faithfully round what is underneath, Fig. 4.2. This may sometimes result in a form that is economic materially, although uneconomic visually in that it demands greater effort in perception and may prejudice the visual unity of the whole, as well as imposing greater problems of holding, machining, finishing and cleaning. This is another example of why aesthetics must be included in any total assessment of an engineering project and why all engineers and engineering designers, the men on the spot,

FORM 55

should be capable of doing this. Trivial economies in material savings can be many times outweighed by loss of visual quality, while the making, handling, finishing and assembly of a complex part must cost more than with a simpler part.

visual unity

Unity and harmony are virtually synonymous and there is clearly some connection between unification and harmonic relationship, although the former is associated more with the reflection of visual characteristics and the latter with steps in size. The link is in the repetition of similar qualities, which can assist identification by making a properly unified visual arrangement easier to perceive and therefore more acceptable than an arrangement without unity. This can be seen clearly with an arrangement of two simple forms, a square inside a triangle, Fig. 4.3a,

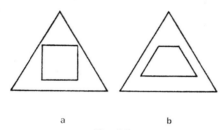

a　　　　　　　　b

Fig. 4.3

where spatially the square seems to attack and to weaken the triangle, a case of the figure attacking the ground. Scanning horizontally, the vertical sides of the rectangle act as stoppers and interfere with the reading of the oblique lines of the triangle, but a slight rearrangement, Fig. 4.3b, keeping the physical areas at about the same values, results in an easier pattern to perceive and to a good measure of visual unification.

Basically, visual unity arises from arranging for some repetition of similar activities for the eye so that the task of the brain in identification is simpler, for having accepted the characteristics of one feature the rest fall more readily into place. It is, of course, desirable that the characteristic being reflected is appropriate in terms of the purpose, function and structure of the form and it is important to realize that consistent quality is more certain to arise from an overall feeling pervading the organizing activity and taking every detail in its stride, than from some very precise and meticulous formula applied separately

56   THE AESTHETICS OF ENGINEERING DESIGN

to each individual feature. This confirms the importance of the strength of general principles and of strong, though perhaps difficult to define feelings, and may make the less tangible aspects of perception easier to understand and to accept.

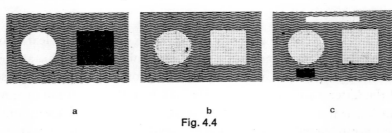

a   b   c
Fig. 4.4

Visual unity depends upon the four basic possibilities of relationship of visual characteristics through colour, nature of surface, shape and size, Fig. 3.13, the easiest of which to achieve is usually through colour. Two forms disparate in shape and colour, say a black square and a white circle appearing on a dark grey ground, Fig. 4.4a, could be immediately related by painting them light grey, Fig. 4.4b. But the strength of that relationship would depend only upon the isolation of

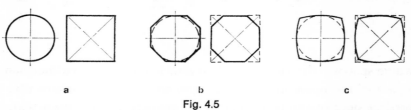

a   b   c
Fig. 4.5

each form from its background and where there are competing spots of contrast provided by other different tones and colours, Fig. 4.4c, the strength of the unification could be seriously challenged. Considerable additional unity could be provided by the repetition of shape characteristics making the circle and the square similar forms, but this would require a more conscious effort and the ability possibly to modify wooden patterns or the basic shape of the two forms, but it would result in stronger visual unity than would be the case with the mere painting of the forms. For example, Fig. 4.5a, the corners of the square could be lopped off and very small areas added to the circle to make both forms into octagons, Fig. 4.5b, or the sides of the square could be curved and the circle compressed or expanded unevenly to make both forms become of quartic shape, Fig. 4.5c, when visual unity could be almost complete.

Once the principle of relating a number of disparate forms by the repetition of visual characteristics has been appreciated it is so self-evident as not to justify numerous specific examples and examination of one may be sufficient. Let us assume it is desirable that there should be some unity between a column and the baseplate upon which it is positioned, Fig. 4.6. Where the section of the column must differ in

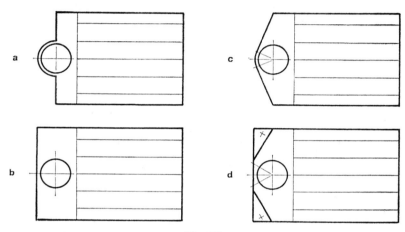

Fig. 4.6

character from the plan of the baseplate, a transitional pad-piece could be introduced upon which the column would rest, but there may be instances, as with radial drilling machines, where the column must be circular in section and the baseplate rectangular and where a transitional pad-piece would be of limited value as it could only be of quartic shape, or a square with curved sides. The alternatives of arranging the base to accept the column as a swelling on one side, Fig. 4.6a, or extending the base to contain the column completely, Fig. 4.6b, result in a somewhat untidy, broken form or in an uncomfortable relationship of the circular column base with the flat end of the baseplate. The cleanest and best-unified result would almost certainly arise from extending the baseplate to contain the column completely, and setting-down parts of the plane of the baseplate area to introduce lines tangential to the column, Fig. 4.6d. This would unify the circle of the column with the rectangle of the baseplate and introduce triangular surfaces at something less than the maximum baseplate thickness which could be suitable for locating holes for holding-down bolts. Of course, the stronger, closest-to-ideal arrangement, would be to retain part of the transitional shape concentric to the column boss, Fig. 4.6c, and to connect it by tangential lines to the rectangular form of the baseplate. But where for economy of material or for any other reason the baseplate length must be kept to the minimum, oblique lines without the concentric link would serve to relate the circle to the rectangle to some degree.

The simplest, and possibly most helpful, thing to appreciate about visual unity is that it is really nothing more than the offering of something of one value to another, and the more that can be offered the greater will be the consistency and the stronger the visual unity.

proportion

Proportion, the relationship of a part to the whole or of one value to another, can be concerned with what is a suitable ratio within a particular context and also with several formal systems having a more universal application. Interest in the latter is not always confined to the solution of an individual problem and there is often an expectation of the revelation of the key to an ultimate truth about form. This has yet to be discovered but it is interesting to consider the function of proportion in engineering design and to reflect upon some of the connections between formal systems and their intriguing links with natural law.

Newcomers to visual matters often do not appreciate that most formal systems of proportion are orthographic and are concerned only with the division of area in one plane. They could perhaps be said to be applicable to the division of volume and space if applied to the three elevations of a solid form, but their general limitation lies in the fact that in perception very few things are viewed orthographically or perpendicular to the line of vision. Almost everything is viewed obliquely and the elevations of objects are usually in several planes and not just in one plane. Thus, the division of flat area on the drawing board need not correspond with what will be seen in the actual hardware and we have to remember that what we are concerned with is not fine theories and formal systems, but with what is actually there in front of our eyes. The equipment bench illustrated, Fig. 4.7, is arranged in its overall

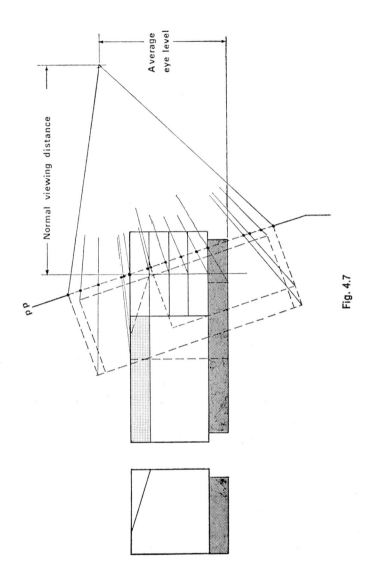

Fig. 4.7

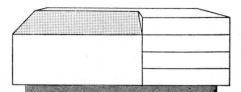

Fig. 4.8

proportion and in its subdivisions, in all three elevations, to comply with the Root Five Rectangle system, and comparison of the orthographic view, Fig. 4.7, with the perspective view, Fig. 4.8, reveals some considerable difference of effect, notably in the areas of the dark plinth and the sloping grey panel. Experienced designers will naturally anticipate some differences of this kind and make the necessary adjustments, but where a principal visual effect is dependent upon orthography, but which becomes something different in actuality, some conscious adjustment of the orthographic image becomes necessary. It is, however, not always appreciated that once adjustment is made to a formal system dependent upon complete accuracy for its working it immediately loses its validity; for example, the Golden Mean system expressed as 1 : 1.617 or 1 : 1.619 instead of 1 : 1.618 just will not work. Too much dependence should not therefore be placed upon formal systems, for although they can be used as a rough basis for the overall planning of size-relationship, with three-dimensional hardware some adjustment is inevitably necessary and once adjustment is made the designer is really on his own. It is much better for him to develop his own sense of relationships, so that he is in control of a situation from the start and can maintain control throughout the various vicissitudes through which the design may have to pass.

Where sound mechanical function and structure are concerned, good proportion in terms of weight and mass, oddly enough, usually gets off to a better start than it does when the principal consideration is of appearance only. This is almost certainly due to the discipline imposed by truth of function, while bad proportion is usually due to some conscious attempt to impose an effect of form which is not a natural concomitant of the circumstances surrounding the design. High-ratio proportions giving distinct emphasis in one direction can be quite satisfactory where they are proper to functional and structural requirements; they are usually self-determining and should present no great

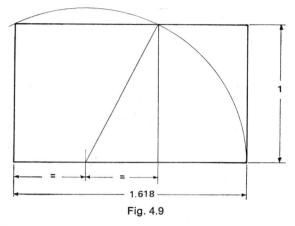

Fig. 4.9

problem. It is in achieving a nicety of relationship and a pleasing degree of variety where values are more equal that problems are more likely to arise. Take the proportion of a simple form, a square, equal in dimensions and in visual tensions and not visually very interesting and consider some of the problems of making it more interesting. Forms with proportions not exceeding 1 : 1¼ are usually still unsatisfactory in that they do not achieve a positive emphasis; they still remain so close to 1 : 1 that they are immediately sensed as being not quite square and this introduces perceptual uneasiness. On the other hand, forms with proportions of 1 : 1¾ and above have acquired positive emphasis and, if this provides a sense of direction appropriate to the application of the forms, there is usually no great problem with them. So, if 1 : 1¼ is the lower limit of emphasis and 1 : 1¾ the upper limit, it may be wondered why has not a midway value of 1 : 1½ become universally adopted as the ideal ratio in this difficult area of proportion? Why have ratios of 1 : 1.618 or even 1 : 2.236 been more favoured? The short answer is that a ratio of 1 : 1½ can be quite satisfactory in some circumstances, but the additional attraction of the Golden Mean, Fig. 4.9, and Root Five Rectangle, Fig. 4.10, systems was that they were infinitely divisible and each subdivision automatically retained the original proportion and was harmonically related to the whole and to all other subdivisions of it. For example, with a large piece of engineering equipment they would, theoretically, provide an overall proportion for the principal structure and properly related sizes for all secondary forms, covers, doors, panels and designation labels. That each system had to be

62   THE AESTHETICS OF ENGINEERING DESIGN

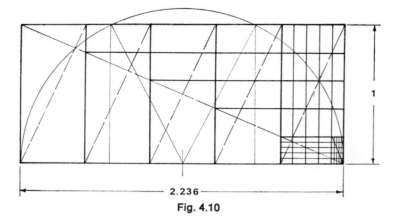

Fig. 4.10

correct to three places of decimals in order to work was offset by the geometric simplicity of their construction, but in trying to employ them generally in engineering design the limitations of orthography applied to three-dimensional hardware soon become apparent. Other ratios in the area of proportion under consideration are 3 : 5 and 5 : 8, belonging to the Summation Series of numbers; their relationships of 1 : 1.666 and 1 : 1.66 are very close to that of the Golden Mean and the connection between these two systems is interesting. The Summation Series was compiled by the Italian mathematician, Fibonacci, in the thirteenth century and the Golden Mean was produced by Greek geometricians about 2,000 B.C. With the Summation Series any number is the sum of the two preceding numbers, e.g. 1,2,3,5,8,13,21,34,55 etc., and after 55,89 the proportion between any two numbers is precisely that of the Golden Mean, 1 : 1.618. Further, if a series of Golden Mean rectangles are built up upon each other and their outside corners connected by a line the result is an Archimedian or logarithmic spiral, Fig. 4.11. Similar spirals are to be found in nature in the arrangement of seeds on the head of a sunflower, leaves in a tightly packed bud, in spiral nebulae and in other natural phenomena. Additional interest in what seems to be the visual expression of some natural law is evident in the work of dowsers and practitioners in psi or extra-sensory perception. Both use a simple form of pendulum to locate the field of forces around a given object and certain rates, expressed as the length of the pendulum cord between the fingers gripping it and the top of the weight, have been accorded to certain basic concepts and found by proof of

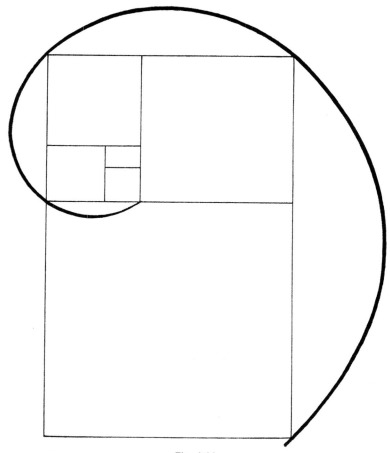

Fig. 4.11

practice to be reliable and, presumably, accurate. These rates are also held to correspond to the radius of the centre section of a biconical force-field surrounding every object, a theory admirably explained by T. C. Lethbridge in his book *A Step in the Dark*. The significance of these rates to the matter of proportion is that if they are set in a circular, compass-rose arrangement, numbering from 1 to 40, and if each spoke in the 'wheel' is marked outwards from the centre with the rates expressed as circles the radii of which are so many 1/40ths of the length of each spoke, and the points at which all of the circles cut the spokes are linked, the result is again an Archimedian spiral, Fig. 4.12. The recurrence of this spiral may be coincidental or it might be

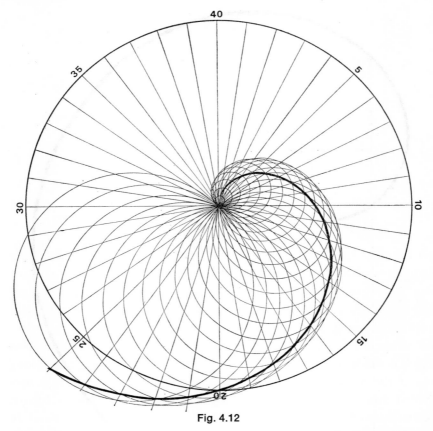

Fig. 4.12

some indication of an ultimate truth about the actuality of objects, but an awareness of such a possibility and the existence of some sensibility about the forms we create can only bring added interest to what can otherwise easily become mundane, boring, everyday work.

The matter of proportion is associated with considerations of balance and composition, and with the achievement of a sense of direction in form.

harmonic relationship

An inherent feature of all formal systems of proportion related in any way to the Archimedian spiral is harmonic relationship, which could be the whole, or part, of some ultimate truth about form. From the illustration of the Root Five Rectangle system the possibility of infinite sub-

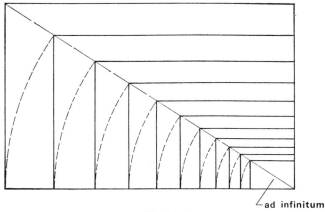
Fig. 4.13

division with harmonic relationship becomes obvious, but if you have modified one of the formal systems or adopted a proportion of your own, how can you obtain harmonically related subdivisions of it? Nothing is easier, Fig. 4.13. You draw out the shape of the basic proportion, put in a controlling diagonal from one corner to the opposite corner, swing the length of the long side of the proportion into this diagonal and where it cuts it you have established the length of the long side of the first subdivision. Repeat this until the whole area is subdivided and you have properly related sizes for any covers, doors, panels, instruction plates and designation labels that may be required. In making the length of the long side of one form become the length of the diagonal of another you have in effect offered something of one value to another and this seems to be the secret of harmony. The basic system of sub-division of D.I.N. paper sizes, Fig. 4.14, is a variant of the controlling diagonal system, from which it derives this particular quality: when each subdivision is folded in half, the proportion of the whole, 1 : 1.414, is retained. Reference to the basic ways of achieving visual unity in the chapters dealing with Perception and Visual Unity should make clear that similarity of some kind is the basis of form-relationship, similarity which can be both in overall shape, colour, texture, size, attitude and in the individual elements of shape. As a rough guide, areas of the proportions of 1 : 1, 1 : 1½, 1 : 2 and 1 : 3 divide into 16, 28, and 80 subdivisions respectively, but these numbers must be subject to some give-and-take because even with forms having a longest side length of 18 in.

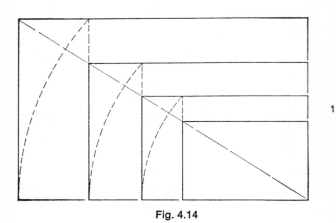

Fig. 4.14

the smaller subdivisions quickly reach a stage where the thickness of the pencil could easily subtract a few. So far as practical, usable areas are concerned those numbers of subdivisions could be taken as 11, 19, 28 and 50, but should the size of the object to which the proportion is applied greatly exceed 18 in. the number of usable subdivisions would come closer to the possible maxima.

secondary forms

Very few pieces of engineering hardware are so simple that there is only a single primary form. There are nearly always secondary forms and these need to be handled with care if they are not to compete too much with the principal form and so weaken it that the perceptual result is unsatisfactory. The chapter dealing with visual unity indicates the basic considerations in relating secondary forms to primary forms and where they are contained within or surrounded by the latter when viewed normally, their integration should not be difficult if the basic facts of form-relationships are observed. But secondary forms which must be realized as physical projections present a greater visual problem. If they are large enough to be perceived as a substantial part of the total form they become attenuations of it, even though they may themselves be symmetrical and not attenuated. If they are not large they may be symmetrical or attenuated, that is, reducing in mass, force or value as they proceed away from the parent form, but the overall effect could

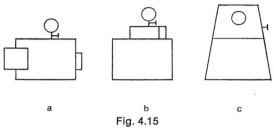

Fig. 4.15

be one of visual attenuation which is somewhat modified by the scale of the form with which it is associated. For example, Fig. 4.15a shows three clearly established projections but their arrangement prevents them from adding up to an attenuation of the principal rectangular form. In Fig. 4.15b however, their relative arrangement now amounts to an attenuation of the principal form and it would be better, perceptually, to integrate them more completely to convert the principal rectangular form into an attenuated polygonal form, as suggested by Fig. 4.15c. With smaller linear forms, Fig. 3.17, such as the strokes of letters, it can result in uncertainty of definition of termination giving rise to a kind of mental hiatus and uncertainty making perception unsatisfactory, a phenomenon examined in greater detail in the section on Sense Impressions. With forms of larger scale, the contour of attenuation is perceived more as the boundary of an area and the question is not so much one of 'Has it finished or hasn't it?' as one of 'What is it?' In such cases it is desirable to achieve a contour as simple as possible to facilitate perception.

Where a build-up of local form outwards from a main form is concerned, one consideration must always be a practical one of structural adequacy, an objective consideration which fortunately usually coincides with the desirable aesthetic and psychological requirements of perception. It is usually desirable to maintain maximum physical connection between progressively extending parts, or to ensure adequate physical connection and support by carrying the extended parts upon a common support, Fig. 4.16. This would logically be of an economic, cantilevered form and how this entirely objective consideration can coincide with general aesthetic aims is illustrated. For economic reasons the various volumes making up any accretion of growth upon a principal form may have to remain as standard forms, as could well be the case with components associated with electricity and other services. In

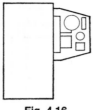

Fig. 4.16

such cases unification is most likely through the introduction of a common cover, as through mounting the parasite forms upon a common support which could itself be of a simple, unifying shape. The obligatory acceptance of forms not aesthetically suited to a situation is a standing problem with engineering design and one aggravated by a tendency towards more rather than less ancillary gear, but not a situation to be accepted meekly without making some endeavour to alleviate the difficulties it imposes.

Generally speaking, an attenuated form is more comfortably accepted where it has a positive commencement and termination, Fig. 4.17,

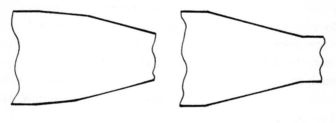

a

Fig. 4.17

particularly where its scale is small. Without thus being defined, visually contained and established, attenuated forms tend to result in perceptual uneasiness, because they have no constant value and any failure to align with the normal perceptual datums may be so slight as to be interpreted more as an accident than as a deliberate intention. Gently attenuated forms of larger scale are usually more acceptable if one of their sides coincides with the horizontal or vertical perceptual datums, Figs. 4.18 and 4.19, for without such visual anchoring polygonal forms can produce perceptual uneasiness. Reference to the illustrations will confirm how some arrangements are aesthetically more acceptable than others and how they can be arrived at through logical, common-sense con-

FORM 69

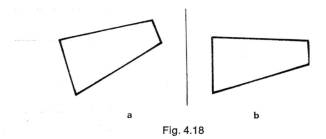

Fig. 4.18

sideration. It is another proof of how objectivity invariably coincides with the general intentions of aesthetes, although the latter tend to cloak their intentions in a shroud of mystique and abhor any connection with anything so mundane as common-sense.

Attenuation means slenderness or rarification; three-dimensionally it means tapering or thinning-off and considerations of the perceptual effects of tapering apply equally well to both rapidly tapering forms and gently tapering forms.

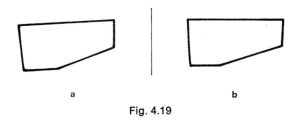

Fig. 4.19

What seems to happen is this. In the case of a rapid attenuation, eye and brain between them have established a datum at which attenuation commences and beyond which it extends, but because what is now accepted as the new form, the attenuated form, needs to be identified and properly related to the parent form, it is desirable that it should be as simple as possible. Where the attenuation is gentle, the taper small, vertical orientation is usually better than horizontal orientation because through perspective diminution we are as accustomed to vertical aberration as we are to horizontal aberration and a vertical form tapered about its centre line does not worry us unduly. But with similar horizontal aberration, Fig. 4.20a, we are more likely to be worried if the lines are not quite horizontal and will be made to feel that the form is not resting securely on a datum, so with horizontal attenuation it is

70 THE AESTHETICS OF ENGINEERING DESIGN

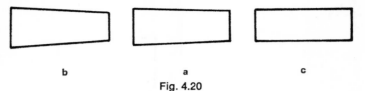

b   a   c
Fig. 4.20

usually better to exaggerate the taper, Fig. 4.20b, or to eliminate it altogether and make the form parallel sided, Fig. 4.20c.

direction

Establishing the direction of a form may seem something of a mystery to some people but, like most other aesthetic manoeuvres, it can be reduced to a matter of common-sense, objective decision.

If we consider a variety of forms ranging from the horizontal to the vertical minima, as is illustrated, Fig. 4.21, and read inwards from either end, we arrive first at points Z, where obvious and unmistakable directional quality ceases and where competition from the opposing elements commences although is not yet sufficiently strong to engender doubt about direction. Moving further inwards we reach points Y, where the competition of the elements is now so great that some conscious effort must be made if the form is to suggest any direction at all. Finally we arrive at point X where the competition of the elements is at its maximum and there is visual stalemate, a completely directionless form.

With engineering products, structure and mechanism can often dictate a near-square format although function and purpose might be better expressed by a vertical or horizontal emphasis and in such cases cosmetic treatment rather than orthopaedic surgery is usually the only practical solution, Fig. 4.22. Such treatment should, however, be qualitative rather than quantitative and any belief that the solution to a problem of visual competition where dimensions and visual tensions are fairly equal lies in an overwhelming preponderance of one set of elements over another is wrong, and it can be self-defeating. The greater the number of the predominant elements the stronger may be the pattern of their seried ends; hence the greater the likelihood of an impression of a square area of tone, where the height equals the width and we are soon back to a directionless, square form. That quality rather than quantity of the predominant element is what counts, and can be seen by reading from left to right in Fig. 4.22. It shows how the desired

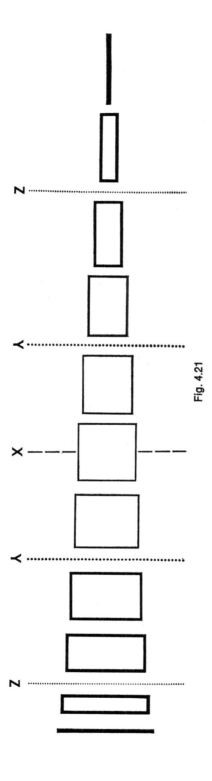

Fig. 4.21

72 THE AESTHETICS OF ENGINEERING DESIGN

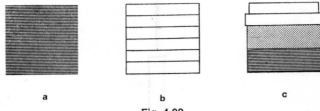

Fig. 4.22

effect can be strengthened by the introduction of small physical changes in dimension, for the visual value of small projecting ledges with their highlights and shadows is usually greater than that of lines of contrasting colour or tone upon a flat surface. But Fig. 4.22 also shows how colour break-up, in conjunction with other contributions, can result in the establishment of some sense of direction in a potentially directionless situation.

Looking again at the illustration of form reading from the horizontal to the vertical minima, Fig. 4.21, those forms between points Y and Z are easier to tip in one direction or another than is the near-square form we have just been considering, but strong competition from the less-dominant elements could equally well restore the balance and leave us with a directionless form. Beyond points Z are forms which would retain their directional emphasis whatever was done to them, although decisive directional effect could be considerably reduced by the visual

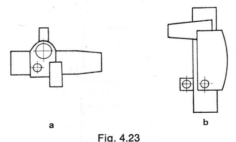

Fig. 4.23

competition from small subsidiary forms or surface displays, Fig. 4.23a and 4.23b. In that case, the restoration of decisive balance is usually only a matter of minor adjustments to dimensions and to tonal or colour effects to allow the principal form to read clearly without distraction or interruption, Figs. 4.24 and 4.25. The fact that in visual competition the answer does not always lie in quantity but in quality shows why there can be no hard-and-fast rules about aesthetic manoeuvre; it also shows that common-sense and objective consideration of the facts of any particular situation can introduce order and make clear a definite intention.

Having looked at some basic considerations of direction in forms, it is relevant to look at some reasons for wanting to give some sense of direction to a form. One reason is to relate it to some functional quality or to a general level or movement of activity with which the form is associated. Another reason is to assist perception of the form, to sort it out visually, to make it decide whether it is supposed to be standing up

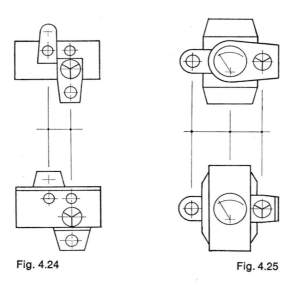

Fig. 4.24   Fig. 4.25

or lying down. Of course, mere physical arrangement to satisfy functional requirement may automatically give some forms direction and there may be no need for any conscious provision of it. But often functional and structural requirements can result in visual stalemate, a facelessness or anonymity making some conscious effort to introduce direction and character essential. The art lies in being able to recognize the point at which self-determination of direction ceases and the need for a helping hand begins; it lies also in being able to spot the essence of the greatest direction-expressing potential and in being able to exploit and emphasize it with the greatest economy. Creating a sense of direction is, of course, closely linked with the competition of visual elements

74 THE AESTHETICS OF ENGINEERING DESIGN

and reference to this in the chapter dealing with Perception would be helpful.

A simple example of the creation of a sense of direction through the alteration of the balance of the basic visual elements can be seen in the illustration of the rolling-stock, Fig. 4.26. The arrangement of detail produces an overwhelming balance in favour of the vertical elements, Fig. 4.26a. Within the context this is not good, because just as the eye tends to fly to or be stopped by a vertical element which can be a welcome relief from the monotony of the horizontal, so a vertical element in movement tends to drag the eye along with it. This could result in a flickering, even stroboscopic, effect which is a halting effect and in the case of the example shown it would be just the reverse of the effect desired. Fig. 4.26b shows the balance of the elements corrected in favour

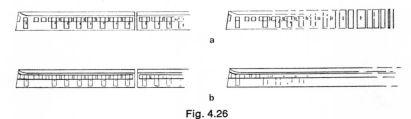

Fig. 4.26

of the horizontal, by means of several practical and permissible devices aimed at reducing the visual value of the vertical elements and then combining a number of reduced-value vertical elements into a single, stronger horizontal element. Such an arrangement would be better within the context, in that just as the eye passes most easily over horizontal elements so does a horizontal element pass most easily in front of the eye, without inducing flicker or drag and suggesting what is appropriate within the context, which is swift, easy movement.

This simple, basic example may seem elementary but this same kind of visual organization can be applied successfully to more complex objects. The aim with the second example, Fig. 4.27, was not solely to establish a sense of direction although that was part of the aim, but this example is relevant to the activity of creating a desired effect and of establishing a sense of certain qualities. With the lathe the permissible work area is horizontal in emphasis and one of the aims was similar to that of the first example. although the intention here was not so much

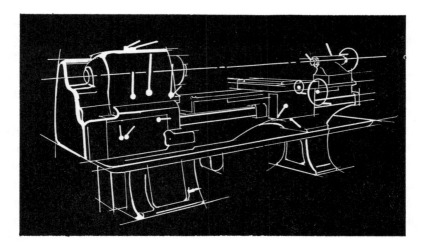

a

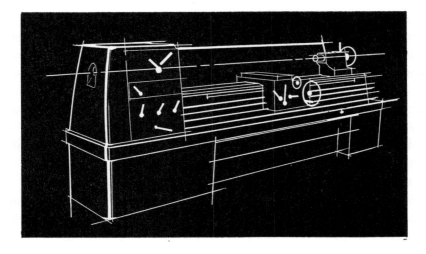

b

Fig. 4.27

to introduce a sense of movement as to suggest a strongly based structure with the maximum, uncluttered work area, Fig. 4.27b. Therefore the competition was not so much a straightforward contest between verticals and horizontals, as between strongly established elements and weak and distracting elements, Fig. 4.27a. But once the principle of numerical balance applied to visual values has been grasped, it can be employed in the solution of the most complex and difficult problems of visual organization. It will soon be appreciated that good strategy can be as effective as massive tactical effort and some notion of what the eye has to do and of what the brain makes of what the eye sees can be of great assistance in determining strategy. We can be persuaded that something is taller and flatter, longer or shorter than it actually is, by exploiting the irrational element in perception and by causing the eye to report to the brain little effects or extensions of activity to which the brain will attach a possibly undue importance. If, at the same time, we are able to manipulate things so that in the opposite direction the eye has little work to do and nothing much to report to the brain, a definite overall sense-impression will emerge and a possibly ambiguous, indefinite situation will have been converted into a positive one. This art of visual organization is not an inspired, heaven-sent gift; it can be cultivated by simple, objective thinking and I emphasize this truth to attack the widespread misbelief that art is Art and engineering is Engineering and ne'er the twain shall meet. In spite of the aesthetically unfortunate results sometimes achieved by engineers, they are usually working to a concept of whatever it is they are designing, that it should be compact, clean-lined, robust-looking or expressive of some other overall quality. That their aesthetic sensibilities may not be sufficiently developed to keep what is produced in line with their concepts does not make them Philistines, for if presented with an aesthetically acceptable solution they are usually able to appreciate it, even though they may be unable at the moment to achieve it themselves.

So important is this kind of organization which can create visual order and achieve clearly-defined form with a positive sense of direction that it is justifiable to emphasize it, to emphasize, too, that it is not difficult to do once the principles of the competition of the basic visual elements are appreciated. For example, if the aim is to create a simple, clearly horizontal form, it is necessary to arrange the balance of the basic linear elements to produce a preponderance of horizontal elements. It is as simple as that. Wherever vertical elements or features contributing to a vertical emphasis exist, they are played down by reducing their visual value or are eliminated altogether. Fig. 4.28 shows a simple example of this kind of manipulation and suggests how visual values can be altered. For example, the value of an obligatory number

FORM 77

of verticals, Fig. 4.28a, in a situation where overall horizontality is desired can be lessened, first by changing those verticals from soft radiused changes of plane with accompanying highlights and tonal gradation, all adding to the visual value, to ones with a smaller corner radius, or to no radius at all, Fig. 4.28b. They could then be further reduced in value by the elimination of the gaps between them, Fig. 4.28c, an obvious step towards making them all of the same height so that they could be unified, Fig. 4.28d; when what were ten strong units of vertical value and ten weak and disconnected units of horizontal value would become two very strong units of horizontal value containing, but

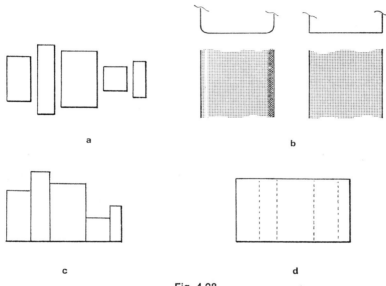

Fig. 4.28

not unduly weakened by, four weak units of vertical value. Indeed, as a single cover would now be possible these verticals could be eliminated. Once the simple basic aim is appreciated numerous opportunities for achieving contributory effects, too subsidiary to be listed in detail, will become apparent. This kind of visual organization can serve to emphasize purpose and function as well as to introduce a more definite aesthetic quality and must result in the creation of fully satisfying forms. Explained in detail, the procedure may seem to be somewhat complex and too consciously contrived, although in practice it may be done simply and without great effort and, as visual thinking becomes a normal mode of thought, with instinctive sureness.

The principal aim of this close-up of aesthetic manoeuvre is to emphasize the truth that aesthetic decision is not necessarily arbitrary nor entirely subjective, that there is a solid basis of objectivity to it and that as aesthetic aims can be achieved by common-sense, reason and logic as surely as by intuition and artistic ability, there is no reason at all why an engineer should not be able to give his work pleasing visual quality. It must be expected that the work of an aesthetic specialist, who need not be an engineer but who must have a good understanding of and respect for engineering, will have a better-unified, more sensitive and perhaps more imaginative quality, but these are qualities expected of a specialist.

The truth about engineering design is that, unless the specialist's field of specialization encompasses more than aesthetics and artistic skill—which is very unlikely and, by definition, impossible—he is little better placed to make a really sound and satisfactory contribution than is an aesthetically insensitive but technically very good engineer. The greatest contribution would, of course, come from an aesthetically sensitive engineer and the enticing and comforting belief that success can be assured by the collaboration of good aesthetics specialists with good engineers is true of only such a minute proportion of the engineering industry and only where happily fortuitous, or rarely fortunate situations obtain, that the real hope for any significant improvement in the visual quality must rest with engineers and engineering designers. That the aesthetic sensibilities of many engineers make this hope appear to be a forlorn one is no reason for abandoning it. The pursuit of truth is never easy and no doubt some time may pass before truth comes into its own, but if we all give up the pursuit it never will.

FIVE

aesthetic manoeuvre

ENGINEERS ARE often at a loss to understand why some effects of positioning and relationship should be considered aesthetically preferable to others, even though they may subconsciously agree that they are. The virtues and the vices of visual arrangements are in fact susceptible to rational assessment and to explanation in terms of mechanical and structural analogy, with which engineers are quite familiar.

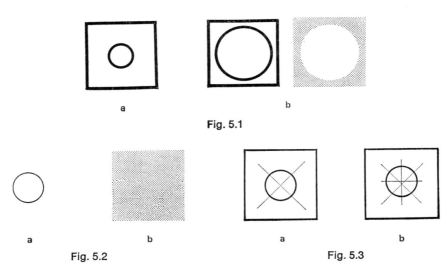

Fig. 5.1

Fig. 5.2

Fig. 5.3

Take the example of the positioning of one simple three-dimensional form upon another which, two-dimensionally, can be seen as a figure on a ground. Assuming that the figure/ground relationship is not likely to be subject to a reversal, Fig. 5.1, we could start with a positional relationship where the figure, a circle, Fig. 5.2a, is placed on the ground, a square, Fig. 5.2b. First, the figure could be placed at the geometric centre of the ground, Fig. 5.3a, next in a preferred position, Fig. 5.3b.

But why should this second position be preferred? There are two reasons why. First, if the first arrangement was viewed at or below eye level, any forward projection of the figure would result in its reading more in the lower part of the ground than in the upper part and it would have appeared to have slipped down, Fig. 5.4. Further, and ignoring any

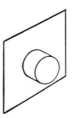

Fig. 5.4

forward projection of the figure, it would seem to be at mid-point of suspension in the frame of the ground, the square, just at the point of losing stability and of slipping down with nothing in hand to ensure it saying where it is. What is postulated as the means of the maintenance of visual stability are chiefly tension links and compression cushions, and the interpretation of a visual situation within the terms of this analogy is not difficult, Fig. 5.5. When the links are not stretched and

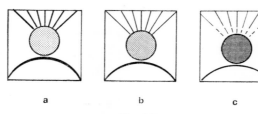

Fig. 5.5

the cushion not compressed, Fig. 5.5a, the situation is a stable one. But if the figure, which can be regarded as a weight, is lowered within the frame of the ground, Fig. 5.5b, the links become stretched while the cushion becomes smaller and loses some of its supporting potential and there is a possibility of a loss of stability. If the figure is then lowered further, Fig. 5.5c, the links become stretched beyond their elastic limit while the cushion becomes fully compressed and the situation is no longer a stable one. Of course, mechanical and functional requirements may demand that the figure be located low down on the ground, the circle low in the square, Fig. 5.6a, and here the best solution could be to introduce a second ground to which the figure would then be related more immediately than it would to the first ground, Figs. 5.6b

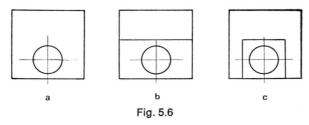

Fig. 5.6

and 5.6c. This new ground could be created by a physical change of plane, by a cover joint line, by colour, tone or some other change in surface characteristic, when the relationship of the figure to the ground would be as illustrated and the composition made a stable one. Mechanical and structural requirements often demand a relationship of forms

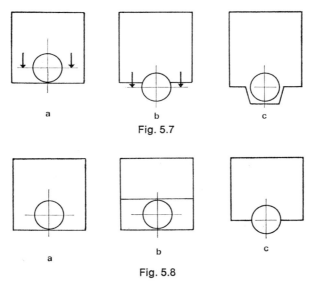

Fig. 5.7

Fig. 5.8

which is not visually satisfactory. For example a circular boss, bearing or outlet, may be required to be located right on or just below the base line of the containing or supporting form, Figs. 5.7 and 5.8. In either case a potential dynamic element, suggested by the arrows, is introduced into the composition. In such cases a satisfactory solution could

82 THE AESTHETICS OF ENGINEERING DESIGN

be to locate the smaller form, the figure, on the base line of the supporting form, the ground, and by adding a small visual extension of the ground, Fig. 5.7c, or by creating a secondary ground or a new structural relationship, Figs. 5.8b and 5.8c, the composition is made stable. The same principles of spatial relationship can be applied to most visual arrangements and it will be found that what may at first have been seen as some quite elusive, mysterious and intangible situation is, after all, quite susceptible to common-sense analysis.

clarity of form

An adequate appreciation of what happens when anything is viewed may be gained from considering the perception of a two-dimensional

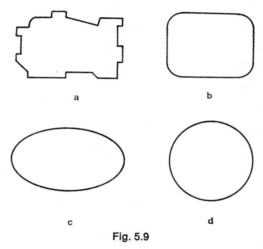

Fig. 5.9

shape, when what happens is somewhat less complicated than with three-dimensional form where stereo depth perception is involved. The tasks of the eye and the brain in perceiving a roughly rectangular form can for our purpose be compared with the tasks of the body and the mind in travelling round an area of similar shape, Fig. 5.9.

If the path is devious with many twists and turns, Fig. 5.9a, the physical effort will be considerable and the memory of the course and the shape it followed will be confused and uncertain. But if the four sides of the area are simplified to create four straight tracks with turns only at each corner, Fig. 6.9b, the physical effort will be considerably less and the memory of the shape of the course will be clearer. If the area was further simplified by curving the four sides and rounding the four corners, the track would have become a continuous one and following it would involve only a variable rate of change of direction, less effort would be required and the memory of the shape of the course would be much clearer. The ultimate simplification would be to reform the area into an ellipse or a circle, Figs. 5.9c and 5.9d, when the track would become a continuous one, with a more, or completely, constant rate of change of direction. The overall effort of following such a track would be but a fraction of what it originally was and simple though this example may be, it may be helpful in the matter of the simplification of visual form. There are, of course, various practical reasons why forms cannot always be reduced to ultimate simplicity, but so long as there is an awareness by engineering designers of the forms they create and of the tasks of perceiving them, there is some chance of them being reduced to a state of simplicity making perception as easy as possible in each set of circumstances.

Taking the two-dimensional shape example a step further, Fig. 5.10, it could be supposed that some of the projections making the original shape so complex were necessary to clear internal mechanisms or movements, Figs. 5.10b, 5.10c and 5.10d. This requirement is likely to be over only part of the thickness of the form, so a third dimension could now be added, and by setting the projection back from the principal face of the supporting form, Fig. 5.10c, it can be read as a complete, simple form: a secondary form attached to a primary form. Further study of the projection might show that the need for it was to clear a lever working only in the upper-quadrant relative to its pivotal centre, and with the original spatial arrangement this need not have been of great significance as the form was already visually complex and uneconomic in space occupied. But with the new arrangement of the form, the projection would have greater significance and it would be worthwhile to examine whether or not the lever would work as well in its lower-quadrant, Fig. 5.10f. It it would, it could be contained within the main form and the need for any projection would disappear. The possibility of often quite fundamental change is not usually attributed to purely aesthetic consideration, but such consideration should be a natural part of any engineering intention and it can often be as significant as any purely technical consideration.

84 THE AESTHETICS OF ENGINEERING DESIGN

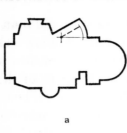
a

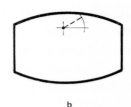
b

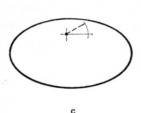
c

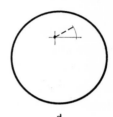
d

e

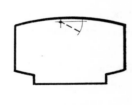
f

Fig. 5.10

Clarification of form is often necessary to establish structural validity and to counter effects arising from any manifestations of irrationality which may be suggesting structural improbability. Querying the need of certain mechanical and structural provisions is usually well worthwhile and can lead to the complete reassessment of a project. It is, of course, not always possible to avoid projections and where it is necessary to have projecting secondary forms, but it is undesirable or uneconomic to enlarge the primary form to contain them, the latter can be clearly-defined to allow it to read over the projections. Eye and brain are thus able to identify and to characterize the principal form without its being prejudiced by the necessary, but usually unwelcome, projections. A typical example of this could be a column with a cantilevered arm projection, Fig. 5.11, which, realized as a single form of uniform thickness is not usually visually satisfactory; the resulting notched-out arm form suggesting a crude hammer shape or heavy-handedness generally, Fig. 5.11a. Although it may well be undesirable for functional reasons to reshape the part radically to make its form more meaningful, a simple and effective correction could be to make a slight adjustment of surface levels, to permit the column to become clearly established as the principal form, with the cantilevered projection as a secondary form, Fig. 5.11b. An analysis of the total function of a form should always clearly establish the function of the parts, when structural sense and meaning are given to them. But while it is usually possible to

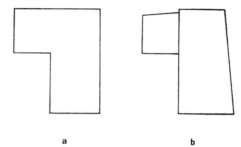

Fig. 5.11

preserve a strong overall sense-impression, even while observing any limitations upon ultimate simplification, it is essential that as much thought be devoted to aesthetic aspects as is given to the more material aspects.

clarity of expression

What is the virtue of clear expression of a visual form? The content of an oration by a 'windy' verbose person may in fact be as adequate as that by a clear, succinct person, but the difference is this. The first speaker has demanded greater effort of understanding from you while at the same time leaving you more susceptible to distraction, but the second speaker has been clearly understood with the minimum of effort and with less possibility of distraction. Had you to appoint a permanent advocate, which is what a product is, you would select the second speaker and not the first. To appreciate clearly the influence of fuss, ambiguity, competition and distraction where visual values are con-

86 THE AESTHETICS OF ENGINEERING DESIGN

cerned it is necessary to understand a little about perception, a word describing the activity of the mind connected directly with activation of the various sense-organs by external causes, and which has already been touched upon.

Clarity of purpose and intention should, of course, be related to the mode of use of the object. For example, the scale, Fig. 5.12a, shows the total arc of movement of the pointer of an instrument, divided into the maximum number of divisions for very precise and accurate indication; such a display would require to be read at normal book-reading distance. But where the precise indication of an exact value is of less importance than a stronger and more clearly displayed indication of a general situation, the scale shown by Fig. 5.12b is better. It permits of a sufficiently accurate estimate to within plus or minus one of the original

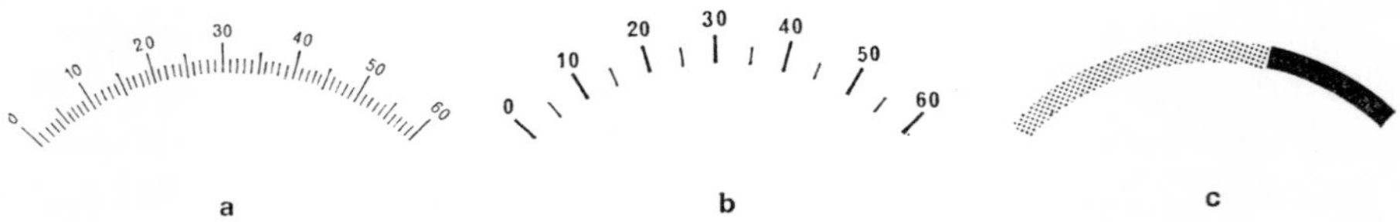

a   b   c

Fig. 5.12

scale divisions, yet could be read at three or four times the distance. Where an indication of a permissible limit is of greater interest even than a rough value, the arrangement shown by Fig. 5.12c could be employed, where the scale has become a plain arc of one colour up to a certain value and changing to another colour beyond it. This indication could be read with certainty at ten or twenty times the distance at which Fig. 5.12a could be read, but it is, of course, only possible where such a mode of information display is acceptable. A further example of clarity of purpose and intention is shown by the direction-indicating arrows, Fig. 5.13. The thin arrow at the top is clear and is analogous to the finer scale in Fig. 5.12, and would demand similar circumstances of reading, for unless it remained within normal book-reading distance, about 18 in., and was free from surrounding visual distraction it would be ineffective. To attempt to improve matters by replacing this arrow by a thicker and more legible line with a shaped end would be self-defeating, for while the mark would certainly become more legible at a much greater distance, the indication of direction is not immediately

AESTHETIC MANOEUVRE 87

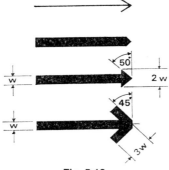

Fig. 5.13

apparent. Indeed, the more equally balanced distribution of interest over the whole mark makes it less indicative than the thinner and lighter mark above. A bolder and clearer mark only becomes more effectual when given a positive directional termination as shown, next below, where the pointed end is twice the width of the stroke and formed at the angle indicated. This is a clear, positive form of direction indication suitable for control panels and instruction plates to be read at distances of up to six feet. The bottom mark, where the stroke width should not be less than one inch, is probably the clearest form of direction indication for reading at distances of ten feet or over.

balance

Visual balance means an equal distribution of interest about an imagined visual centre of gravity, Fig. 5.14a. It does not have to be static balance which is symmetry; it can equally well be dynamic balance which is asymmetry, Figs. 5.14b and 5.14c. The difference is this. In the first

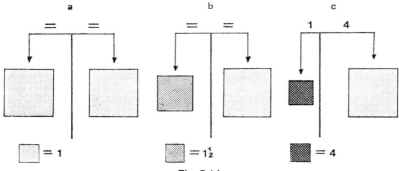

Fig. 5.14

case, Fig. 5.14a, there could be two visual values of the same weight or mass arranged equally about a vertical axis or pivot point. In the second case, Fig. 5.14c, there could be a small value and a large value and the visual axis or pivot point would require to be established where it provided the appropriate leverages to result in balancing moments. Where the visual value of the two elements is similar, in that they are of the same quality but vary in quantity, the visual axis or pivot point would clearly need to be off-centre, or asymmetric. Wherever the axis or pivot point needs to be centrally placed, or to be located in any particular place because the visual circumstances demand it, balance must be achieved by adjusting the visual value of the two elements involved to suit, Fig. 5.14b. For example, a small, concentrated weight can balance a large, diffuse weight where both have similar leverage, just as a small piece of metal might balance a large piece of wood, which raises the question of how can visual weights be adjusted.

They can be altered by emphasizing or diluting tonal contrast, by strengthening or lessening the visual value of any containing frame or structural feature and by doing anything else which will, in the circumstances, add to or subtract from visual importance. Once the habit of visual thinking has been developed, visual weights become automatically ascribed, some stimuli being rated as lead while other stimuli might be rated as feathers.

Placing, or the balancing of the individual elements of a visual composition, has always been conceded as being an art and due recognition accorded to those with the gift for it, but while an ability to achieve nice visual balance may well be intuitive with some people, there is nothing mysterious about it, nothing that cannot be attained by an objective approach to any problem of arranging a number of visual elements to produce an agreeable and reassuring effect.

Decision upon the location of any visual axis or pivot point must be influenced by other visual features in or adjacent to the group of elements to be balanced, Fig. 5.15. There may for example, be dominant lines created by covers or by changes of plane in the background which will automatically create a visual axis, about which the other elements must be balanced. This is where the elements to be balanced must be adjusted to the situation.

Where the elements to be balanced are isolated on a ground free of

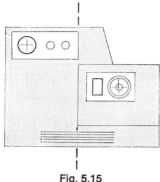

Fig. 5.15

any dominating visual features the axis or pivot point can be established to suit the elements. This is where the situation would be adjusted to the elements to be balanced. Eye and brain with their uncanny ability to sense the position of an axis or pivot point will usually do that automatically and where there are visual features already established at, or close to, the imagined position they will either reinforce or conflict with the line of the axis which is felt but not seen. There may therefore sometimes be a need to adjust all elements within a given area in order to achieve an harmonious relationship. The effect of the chain-reaction of visual elements is not always appreciated by those unfamiliar with visual values and changes are often made which either make the changed item no longer consistent, or which make many visual features not directly connected with the changed part no longer valid. Appreciation of what are the basic relationships enables unity based upon one characteristic to be strengthened by increasing the linking to two, three or four characteristics. The disruptive effect of the assembly of many engineering forms, particularly the effect of many accretions of minor forms, is due to an unnecessary variety of visual characteristics. This, in turn, is usually due to the individual forms having been designed in their own right and not as visually integrated parts of a larger form, which of course happens in the case of many components and accessories that are standard, proprietary articles. With these one may be restricted to one link of relationship, colour, but where shape and size are also able to be exploited the relationship can be much stronger and what might have been a quite disparate assembly of forms could become a very well related assembly.

symmetric or asymmetric form?

Whether a form should be symmetric or asymmetric should never be a matter for purely stylistic, emotive or subjective decision, nor should it ever be a matter of dogmatic opinion. There should thus never be any problem, because the requirements of the design should always indicate clearly what should be done in any particular circumstance.

Symmetry can be perfectly suitable, although there is not the slightest doubt that it is too often selected as a safe, satisfactory solution and imposed upon situations where in fact asymmetry would have been a better decision. Where symmetry would mean adding unnecessary structure merely to achieve static rather than dynamic visual balance, or where it would make proper sequential arrangement impossible,

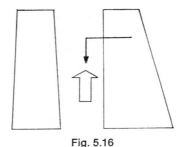

Fig. 5.16

asymmetry is the correct decision. It is however rarely justifiable where it is used merely for subjective reasons and certainly never justified where it results in an unequal, unbalanced, distribution of structural loading.

A simple introduction to the subject is through a column form, Fig. 5.16. Columns subject to lateral thrust, as are many column forms in engineering design, may be considered as vertical brackets and the only economic form for a bracket is a tapering one or one with a reducing section. The front elevation of such a column may be symmetrical, but in the side elevation it could well be buttressed to withstand upward pressure on the front face and thus be wider at the base than at the top. If the front face has to be kept vertical, this structural reinforcement would result in an asymmetric form. All C frame structures must be asymmetric in one plane at least or they could not fulfil their function and if the simple and obvious reasons for why this is so are appreciated there need never be any doubt as to whether a form should be symmetric or asymmetric.

Asymmetry is perfectly justifiable where mechanical function or access for work entry demands it and it often fits in more naturally with the total work or operational relationship between a human being and an engineering product than would symmetry. Symmetry is balanced in a static way and tends therefore to be more self-contained and less well-integrated with the system of which the product is but a part.

Of course, a situation may occur where a symmetrical form would be most suitable, but where the surface display on the form requires to be asymmetric for reasons of proper sequential operation. The best procedure in such a situation is considered in the next chapter.

SIX

surface treatment

THE TREATMENT of a surface is almost as important as its physical development. It can be embellished by things applied to it locally as controls, covers, doors, fastenings, instruction and name plates, or generally as organic or inorganic coatings. Aesthetically, these two modes of application can be generalized as graphic display and finish.

Graphic display covers all information on the man/machine interface and information in this context means any kind of communication between the product and the operator. As communication can be tactile as well as visual, physical controls such as buttons, switches, levers, knobs, handwheels and items such as scales, meters and gauges may be classed, with letters and numerals, as information within the context of graphic layout.

Finish covers colour and texture and concerns paint, plastic and metal films as well as chemical or physical rearrangement of the surface.

letters and numerals

All letters and numerals are just marks, figures within the aesthetic context, but figures with special significance. European and Russian communication systems employ no more than thirty letters and the need of that number of readily-recognizable variations of mark points to a principal difficulty with any language. The possible permutation of the pattern of the mark ensuring adequate difference and clarity in perception is limited, while the number of letter faces affording especial clarity is further limited. In addition to aiming at clarity, all type and letter systems are subject to some degree of style and fashion and may also be conditioned by purely material and technical requirements of reproduction, but the outstanding quality of the best systems is their clarity, the ease with which they can be understood; a simple virtue applying to other forms of mark such as symbols, or code languages, used where graphic analogy would be more succinct than would be a number of words. The test for suitability is the same for letters and numeral forms: it is not how theoretically or intellectually appropriate they may be, but how easy they are fully to perceive, identify, characterize and understand; back again to the importance of what is actually in front of our eyes and of what we construct from what we see.

The principal aim of a typographic specialist is to achieve clarity of presentation, which is far more important than compliance with any current fashion or convention. Nicety of spacing, fine balance, achievement of tone, distribution of weight, in fact most specialist typographic aims, are intended to make the task of perception easier by presenting the eye with a more constant loading and the brain with a more even

SPACING
a
SPACING
b

Fig. 6.1

task of recognition and understanding. The eye should not for one minute be confused by a number of vertical strokes too closely spaced for clarity, Fig. 6.1, and the next minute be confronted by an unduly open, empty space. Evenness of weight, that is, of figure and ground, produces constancy of demand upon the optical system facilitating reading and, coupled with the more immediate recognition of what is read, results in full perception. A great deal of typographic expertise can thus be seen to be little more than common sense made visible, to borrow the late Frank Pick's definition of design.

The choice of a particular typeface should be influenced by the overall requirement and should not be an arbitrary, fashionable or stylistic choice. Where phrases of many words are involved it is undesirable for them to appear in capitals only: these are better reserved for isolated headings, functional designations and mandatory instructions. Psychologically, capitals are more impersonal than upper-and-

94 THE AESTHETICS OF ENGINEERING DESIGN

lower case lettering, being associated no doubt with street nameplates, newspaper placards and headlines and bald, official signs. Upper-and-lower case lettering is more personal, being associated more with handwriting, from which of course it developed; therefore it is a more intimate form of communication and more appropriate, and effectual, for directions and instructions. Within a certain permissible measure,

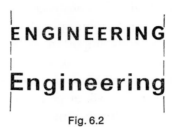

Fig. 6.2

that is, length of line, upper-and-lower-case letters can be more immediately legible than was the same information presented in upper case only, Fig. 6.2. This is yet another example of how maximum visual effect is dependent more upon quality than upon quantity; indeed, as the weight and boldness of the letters increase, without any extension of the measure, or the line length, the less legible the word becomes. Any chosen face will thus almost certainly need to be available in small, lower case, letters as well as in large, upper case, letters. This

LIGHT

MEDIUM

BOLD

EXTRA BOLD
Fig. 6.3

terminology originated in the fact that printers' type for handsetting is normally arranged in tiered trays or cases from which the compositor selects the metal type slugs to arrange in his 'stick' or hand-held assembly frame; the large letters are kept in the upper case and the small letters in the lower case. In addition to being available in lower as well as in upper case the face may need to be available in a number of weights or degrees of boldness, such as light, medium, bold and extra bold, Fig. 6.3. This may be necessary to be able to deal with a variety of displayed messages and also to enable the introduction of some degree of visual priority other than by variation in size, not always practicable. So where it is desirable to emphasize a particularly important part of what is being displayed, variation in weight will usually achieve this better than will variation of size or colour. It might also be desirable to present particularly important or dynamic commands, instructions or parts of a message in italics, a form of lettering introduced into England by a German printer early in the sixteenth century, in which case the chosen face should also be available in that form, Fig. 6.4. Of

ITALIC fount

Fig. 6.4

course not all type faces have a full range of weights or are available in italic form and to have to resort to employing a mixture of faces upon one job is to surrender some visual consistency, the maintenance of which is not only desirable aesthetically but is also conducive to more satisfactory perception. So here again, common-sense and the recognition of material and functional facts are reliable guides in what is to many a no-man's-land of arbitrary or divinely-inspired decision.

Whether to employ a face with or without a serif can best be decided

Fig. 6.5

by examining the function of a serif, another example of the questing, why?, why? technique employed so fruitfully by Leonardo. A serif is that little cross-line or swelling which finishes off the ends of strokes, or limbs, of some forms of lettering; arising perhaps out of a conscious imitation of incised lettering where it provides a positive, visible termination of a stroke, Fig. 6.5. Certainly with metal type it provides support for what might otherwise be thin, upstanding ribs of metal, while there are other virtues of the serif which should be recognized. For small lettering, typical of books, newspapers etc., where the lower-case rarely exceeds ⅛ in. in height, serifs contribute a horizontal quality to an assembly of short and predominantly vertical elements. Thus, to an assembly of lines of lettering, Fig. 6.6, serifs supply a measure of continuity that is consistent with the quality of calligraphy from which printing type was developed; an invention usually credited

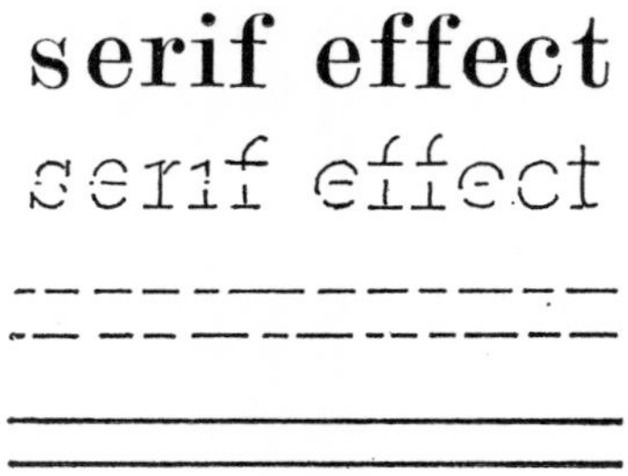

serif effect
serif effect

‒‒ ‒‒ ‒‒ ‒‒ ‒‒ ‒‒
‒‒ ‒‒ ‒‒ ‒‒ ‒‒ ‒‒

―――――――――――――

Fig. 6.6

to Johannes Gutenberg who, in Strasbourg about 1440, established the basic techniques of modern printing. He had however been anticipated by a Chinese smith who produced type of burnt clay about 400 years earlier and also, it is believed, by Koreans who were using copper type about 40 years earlier still. Another function of a serif is to mitigate the rather disconnected, staccato quality that a mass of small sans serif

sans-serif effect
sans-serif effect

ccnilcril clhcill

| | | | | | | | | | | | | | |

Fig. 6.7

type can produce, Fig. 6.7. But in the context of engineering applications their use for bold displayed information has diminished in recent years and sans-serif faces such as Gill, Grotesque, Folio, Futura, Univers and Venus are more commonplace; a trend which is not so much a matter of fashion as a recognition of the functional and perceptual qualities of lettering. To many people, Roman lettering is classical, thick-and-thin stroke lettering with sharp, fillet radiused serifs, Fig. 6.8, reminiscent perhaps of the lettering on the Trajan Column, old buildings, manuscripts and tombstones. In typographical terminology, however, Roman describes lettering of any style, with serifs or without, that is upright; lettering that is not upright is italic. So when a typographer specifies Roman lettering he is thinking only of upright letter-

ROMAN
a

ROMAN
b

Fig. 6.8

ing, which might well be in the most modern sans-serif style and not necessarily in any classical style.

Serifs are not always those spiky, dagger-shaped things. There are also what is known as slab serifs, Fig. 6.9, which are associated with a series of typefaces classified as Egyptian although mainly of European origin

DESIGN

Fig. 6.9

and the subject of a typographic renaissance in Victorian times. Most of the Egyptian letter forms have strokes of equal or near-equal thickness and closely approach the general character of most sans-serif faces, while slab serifs, as the name suggests, are rectangular in form. They are in fact just a short length of a stroke arranged cross-wise at the end of a stroke to make a more obvious termination of it and to create horizontal lines of continuity helping to even-up round and purely linear letters. Within the context of a generous area of typesetting, serifs tend to make for a more clearly-defined figure and for a more evenly sustained demand upon the mechanism of perception, for where the area of typesetting is considerable, as with the page of a book, the sustained task of the perceptual mechanism is principally that of repeated horizontal scanning. This is clearly best accomplished where there is horizontal continuity, where the only interruptions are deliberate breaks connected with the content of the printed words and as there is a preponderance of vertical elements in lettering, serifs are an aid in relating these to the datums which are lines in which the letters are arranged. There is here a reference back to the basic linear elements and to the visual competition inherent within them. Serifs are horizontal in character and the quality they introduce is in support of horizontality which is perceptually the norm and tends to be uneventful, restful and even soporific, Fig. 6.6. Vertical elements, on the other hand, being variations of the norm, tend to be arresting and attention-holding, Fig. 6.7, the reason why, for short messages, warnings or commands, a sans-serif face is usually more effectual. The only qualification is that the line length must not be too great nor the total area of setting too large, or this arresting quality could become itself an impediment to smooth reading and to immediate perception and thus defeat its own end.

DATUM

Fig. 6.10

There is, of course, no suggestion here that serifs are *passé* and should not be used, merely a suggestion that for notices of the general character and setting area that engineering is most concerned with, a sans-serif face is generally better. The forms of sans-serif faces are legible and more consistent with the logical, geometric forms of engineering than are the more 'bookish' qualities of serif faces. However, a quality possessed more strongly by serif than by sans-serif faces is a suggestion of refinement and hair-line accuracy, Fig. 6.10, and if the product concerned is an instrument or a piece of high-precision equipment a serif face could be quite appropriate for a displayed name or for messages which do not have great urgency. For many domestic products too, a serif face could be appropriate, with the same reservation about urgent or vitally important pieces of information.

A further consideration in selecting a style of type concerns numerals, which in engineering applications of typography are employed almost as much as letters, and not all typefaces have really legible, unambiguous

SURFACE TREATMENT 99

numerals, Fig. 6.11. It is essential that letters and numerals are consistent in style and ensuring the proper quality for the numerals could sometimes be a justifiable reason for deciding upon the style of the letters. At a practical level it is very useful if the selected face is readily available in a film-printed, self-adhesive, touch-down lettering, which greatly facilitates the production of the necessary 'masters' for the artwork which must be supplied to the plate maker.

In addition to the purely graphic arrangement of any information there is great opportunity for aesthetic quality in the selection and arrangement of the content of what is displayed, which is often poor with an insensitive, inappropriate choice of words and weak sequential arrangement which fails to convey a clear, concise intention. A graph,

1234567890

1234567890

1234567890

1234567890

a

1234567890

b

Fig. 6.11

simple figure or diagram based upon the sense, connection and consequence of the content can often be a great help in final selection and arrangement. Such graphic expression of mental processes is possible and could no doubt be used to greater advantage than it is; it is not finite or measurable in any way, but just as graphic expression of rationalization and optimization can reflect the condition of what is under review so can it reflect a strong, weak or ambiguous intention with information. Nothing more than competent doodling is required, but this kind of visual thinking can often achieve results matching those of an entirely mental approach or it can provide a reassuring check upon results so obtained. The development of an ability to think visually is of inestimable value to all connected with design and is, of course, of special significance in the development of aesthetic sensibility.

layout and balance

All information layout should be on the principle of first things first and this should be reflected in the visual priorities accorded. First, what is being looked at should be clearly identifiable. Second, where necessary, the occasion of use should be made clear. Third, where necessary, the mode of use of the information should be made obvious. Brevity and clarity are two cardinal principles and, with correct sequential arrangement, should ensure good communication. The shape of any controls, be they buttons, knobs, lever-endings or handwheels, should indicate their function and suggest how they are intended to be used,

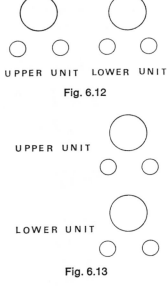

UPPER UNIT   LOWER UNIT

Fig. 6.12

UPPER UNIT

LOWER UNIT

Fig. 6.13

and should be comfortably sited and suited to the best mode of operation.

The arrangement of displayed information should so far as is possible reflect an actual state of affairs or anticipate resultant movement. For example, controls and indications relative to two sets of elements situated one above the other, an Upper and Lower situation, should not be displayed on the same level, which is a Right and Left situation, Fig. 6.12; they would be better arranged as Fig. 6.13. Similarly, the movement of any control should produce an expected and predictable result, a lever moved to the right or left or up and down should result in movement or some activity in the same direction. A handwheel, Fig. 6.14, turned to the right, should result in movement or activity to the right, which is contrary to the result of turning a wheel with gear attached against a fixed rack and will usually justify the introduction of an idler gear to secure coincidence and eliminate hazard, Fig. 6.15. Such a gear may be omitted for reasons of economy although economies of this kind may well be in the penny-wise pound-foolish category, ignoring the truth that the collective effect of the refinement and civilized presentation of a number of small details can determine an overall characterization as either Rolls-Royce or Ford.

All information is an important component of the cosmetic quality of engineering design and can be more important in determining overall aesthetic quality than, it appears, is often imagined. Plates of different kinds are still too often apparently regarded as irritating but necessary

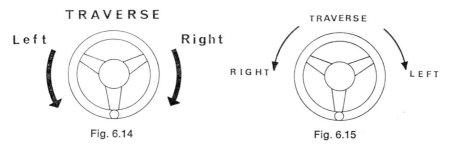

Fig. 6.14   Fig. 6.15

afterthoughts and their placing and attachment left to some very humble member of the works staff, far removed from the disciplinary influence of the design department. As aesthetic quality is our reaction to what we see in front of us, plates-and-all, engineering designers should, in consultation with the appropriate product operation specialists, specify the content of any information, its display, its layout and position and the materials and processes to be used in producing the various items. In this way, and only in this way, is real overall aesthetic consistency possible, and arbitrary or indifferent treatment of a most important element of engineering aesthetic avoidable.

Whether layout is symmetric or asymmetric should be decided entirely by what is best for the legible and meaningful display of the information. Symmetric layout appears to commend itself more readily to many people as a safe, sure, balanced approach; they feel that they cannot go wrong but symmetry, which is static balance, can often be wrong in that it can conflict with the achievement of maximum display value and with proper sequential arrangement. What is being displayed may often lend itself better to dynamic rather than static balance and there should be no preconceived solution; this should only be decided upon by the nature of what has to be balanced. If, for example, there are two controls with only one indication it does not follow that they would be best arranged as Fig. 6.16a; indeed, both controls may need to be adjusted before the necessary indicated reading is determined, when arrangement Fig. 6.16b would be more suitable, or arrangement Fig. 6.16c which is symmetrical on the vertical axis only. Visual balancing is very similar in principle to physical balancing in that the same factors

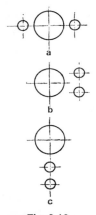

Fig. 6.16

of weight and leverage apply, the only difference being that the weights are visual weights and the levers are visual axes. Just as the value of a physical weight needs to be adjusted to suit what leverage can be given to it to achieve a certain moment, so can visual weights be adjusted. Spotting a visual axis and relating it to some other visual characteristics of the whole composition may not be quite so simple, but experiments with balancing a number of disparate shapes will soon develop sureness and an increased understanding of what it is best to do in certain circumstances.

Where an asymmetric surface layout is essential but where the form upon which it must be displayed is best symmetrical, can present an interesting though not too difficult problem, Fig. 6.17. In such a situation scale has a definite bearing and where the display is small enough to be read on its own and not as part of the overall supporting form, there should be no conflict. The brain tends automatically to relate scale with possible conflict, but where there is a reasonable area of space around one form isolating it from the outline of another form there is no visual conflict. The brain has, in such a case, recognized that the display and the form upon which it appears are two separate things. But where the elements of the display are sufficiently large to be read as part of the total perception of the supporting form, display and form are seen as one set of visual elements and there is possibility of visual imbalance and unease.

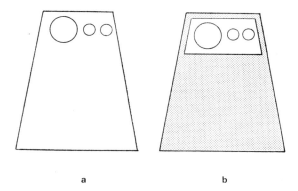

a b

Fig. 6.17

If three elements are concerned, one large and two small, and if the two smaller elements were required to be read or handled before the larger one the overall effect of a symmetrical arrangement would not be satisfactory. Of course the display could be kept symmetrical by placing the two smaller elements centrally above or below the larger one, Fig. 6.16c, and preserving the possibility of a measure of sequential layout, but this would only be suitable where the small elements were associated with one function and the larger one with another. Where all three were associated with the same function it would be better to have a straight, left-to-right, sequential arrangement, Fig. 6.17. Assuming this, the first thing would be to check whether or not the supporting form must be symmetrical, or whether it could be altered to an arrangement upon which an asymmetrical display might sit more happily.

Assuming that symmetry was judged to be best for the supporting form, the next step might be to introduce a secondary form to which the display could be related, Fig. 6.17b, rather than to the primary form; equivalent to a reduction of the scale of the display in that an isolating feature has been introduced. This feature could be made visually more important by colour, tonal contrast or by physical projection or recession and it would permit the asymmetric display to sit happily within the symmetric form. Asymmetric arrangement of a display can, of course, always be given dynamic balance by sensible arrangement of the figures on the ground, not forgetting that words, be they designations, instructions or names, can be used to balance arrangements of numerals.

nicety of display

The colour choice with the majority of standard components and materials is black, white and grey and it is often a disappointment that manufacturing units which consume quantities sufficient to justify colours suiting their own special requirements do not avail themselves more than they do of the opportunity of securing a unique and more consistent quality for their products. Meter and gauge dials, though sometimes black, are normally white with scale markings that are in many cases unnecessarily fussy, Fig. 5.12a. This arises from catering for the user requiring the greatest number of fine markings although in many cases an exact indicated value is of no interest; what is of interest are permissible value limits. Thus, a visually fussy, fine scale could often be a simpler affair with fewer, bolder markings, Fig. 5.12b, or an even simpler and bolder two-coloured arc, Fig. 5.12c. Provision of standard instruments with either fine or coarse scales should not present a great problem. Neither should it be beyond the wit and ingenuity of instrument makers to devise a system of plain-arc scales which could be assembled to suit a number of commonly-experienced permissible value limits.

These considerations may appear to be more ergonomic than aesthetic ones but in terms of visual display there is much in common. A black scale on a white dial is a reasonable standard solution and where meters and gauges have to be seen in isolation and must provide their own insularity the sense of this can be appreciated. But a number of white square or circular 'eyes' appearing upon a medium or dark-toned form can present visual problems. In order to indicate the need of mental vision and imagination in the approach to aesthetics rather than to suggest a realizable, practical solution, imagine a horizontal row of four meters or gauges, Fig. 6.18. The first consideration would be one of legibility, which could be due not only to figure/ground contrast but also to figure clarity, the visual strength of the figure, in this case the scale markings. A black/white contrast amid an area of medium tone

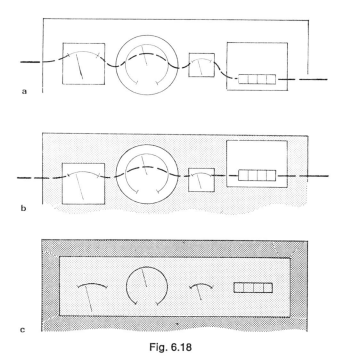

Fig. 6.18

could be too great for immediate perception and thus be self-defeating, particularly where the scale of the figure is very fine and a bolder figure on a less contrasted ground would have equal or better legibility while the ground would be less disruptive of the form upon which it appears. Further, the strength and the visual prominence of a group of white dials situated close together would compel their rigid alignment, although the centre of visual activity of each need not coincide with the geometric centre, Fig. 6.18a. Thus a situation is possible where attention, having been compelled by the visual strength of contrast and alignment, would be obliged to follow an obstacle-race course to perceive in detail the information to which it had been directed. Aligning the dials on their centres of visual interest would, of course, be undesirable without some modification of figure/ground contrast, but given such modification and with the addition of a unifying area of the same tone and colour as the dials, Figs. 6.18b and 6.18c, a result would be possible where the information areas were sufficiently identified and did not disrupt the form upon which they were carried. They would be arranged in an orderly, less jumpy pattern and while initially there would almost certainly be difficulties with meter frames and bezels, once merit was established and demanded, suppliers would soon find ways of catering for it. It must be repeated that these considerations are no more than an indication of a general mode of thought, although it should be appreciated that in some matters, particularly in the incorporation of elements not ideally suited to particular situations, we are still at a fairly elementary stage and elegant, imaginative thinking can do nothing but good.

size

Decision upon the size of any displayed letters or numerals is another design consideration dependent more upon objective analysis and common-sense than upon any particular graphic skill. The requirement is that whatever is displayed should be comfortably legible from the normal viewing distance which, for all functional instructions and information concerned with operation, is from the position the operator must occupy when reading anything relative to the operation. This position may be a seated or a standing one and adequate surfaces for carrying whatever is displayed may be as close as the page of a newspaper held in the normal reading position, or they may be several feet away and analogous more with tickets or showcards in a shop window. But whatever the basic general viewing distance is, it should determine the general size of letters and numerals, which may be increased to maintain uniform legibility the farther the matter must be from the fixed reading position, Fig. 6.19. This, of course, only applies to instructions and information to which the operator must refer during the functional cycle and information relative to setting-up, servicing and maintenance, which need not necessarily be visible from the operating position, is

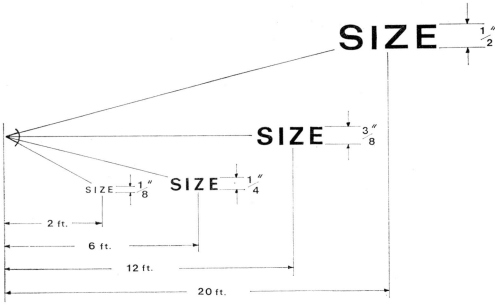

Fig. 6.19

more likely to become standardized in size to what is suitable for an average viewing distance of 12–18 in. A good general rule is for a minimum size of 12-pt., approximately ⅛ in. high for capitals and numerals and $\tfrac{3}{32}$ in. for lower case, for anything viewed at normal viewing distance. For information viewed at something greater than normal viewing distance, for example, the speed and feed plates on saddles and spindle frames likely to move about relative to the operator, a minimum of 36-pt., approximately ⅜ in. high for capitals and numerals and ¼ in. high for lower case letters, is a good general rule. But all decisions upon size should be finally conditioned by individual circumstances and so long as the simple rule of legibility in action is remembered there should be no real problems, certainly no need for anything more than good common-sense.

The over-riding consideration about size is the degree of legibility, which is related only partly to size, partly to clarity of mark, and partly to weight of mark. By merely looking at a variety of type faces this is immediately evident, but is particularly emphasized by comparing the suggested marks in B.S. 3641, *Symbols for Machine Tool Indicator Plates*,*

* British Standards Institution, 1963.

Fig. 6.20, with the M.T.I.R.A. Research Report No. 10 reviewing that standard. For a given size the weak, scratchy symbols suggested in B.S. 3641 have but a fraction of the legibility of the bolder, more positive reassessment of them in the M.T.I.R.A. report, Fig. 6.21. The letters and marks suggested in the latter would be still more legible and more meaningful were they half the recommended size and it is well to appreciate why this is so and to realize that with displayed information quality is more important than quantity, a truth which recurs again and again in consideration of visual effect and manoeuvre.

A study of B.S.3641 makes it very clear that neither the B.S.I. Mechanical Engineering Standards Committee nor any of the thirty-seven co-operating bodies responsible for this standard had any concern for graphic quality, yet the specification of standard symbols, a sign-

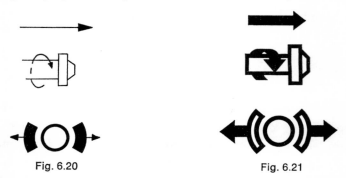

Fig. 6.20   Fig. 6.21

language, is very much a visual matter. To attempt to do it without graphic guidance and assistance was scarcely intelligent and the fact that a major organization within the industry to which the standard was directed had immediately to publish a major assessment of it is sufficient comment; it is yet another example of how low visual quality is rated by many senior organizations which should know better.

colour

Colour is clearly a very important component in the cosmetic quality of the engineering design aesthetic, but it is rather less susceptible to a wholly objective approach and is rather more susceptible to emotion and to subjectivity than is form. It can be rationalized up to a point, in terms of reflectivity, the amount of light absorbed or passed on, of tonal contrast with any workpiece or work environment, and through the avoidance of any sharp tonal contrasts within the work or operating area. This degree of objectivity should indicate a certain range of colours suitable for any given application and it will have narrowed-down the possible choice of hue, value and chroma, the three identifiable variables of colour, Fig. 6.22. These terms vary slightly with different colour systems, but the ones given are those used in the Munsell system upon which the B.S.2660 range of colours is based and therefore the ones in more general use. Hue is the attribute of a colour determining its name

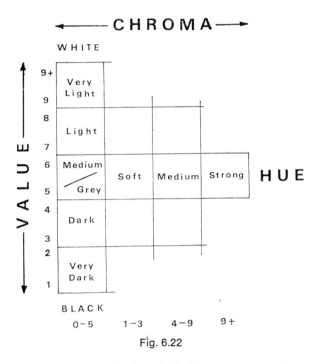

Fig. 6.22

in general everyday use, whether it is blue, green, yellow, red, or of some definite class. Value is the lightness or darkness of the colour, or its tone. Chroma is the intensity of hue in the tone which can, of course, also be influenced by the amount of black, or absence of light. Of the narrowed-down selection of colours, some would almost certainly have been adopted by and possibly established as house-colours by competitors and would no doubt be eliminated. Of the remaining equally suitable colours current trends and fashions may eliminate a few, leaving a residue from which final selection could be made and it is important to realize that this selection can only be on an emotional basis and could reflect not only the personal taste of the designer or the maker but of the wives or girl-friends of either. It is unrealistic to believe that colour selection can be made more finite or positive, but it is usually sufficiently precise to avoid a clinically unsuitable or sensuously arbitrary choice.

The selected colour could, of course, be achromatic, that is, without definite hue, and there is a wide range of chromatic greys—greys with varying blue, green, yellow or brown tendencies—which are very suitable for engineering applications. Their availability in finishes with the right chemical and mechanical properties may certainly need to be improved, but here the engineering industry is probably more to blame than are the paint manufacturers. In Britain we have tended to favour a somewhat limited range of greys, characterized by Light and Dark Battleship, while users of colour on the Continent have always favoured more subtle variations and reference to the D.I.N. Colour Dictionary will reveal a much wider range of standard mixes than is provided by any British standard system.

Colour finishes can be provided by pigmented coatings, or by dyeing or chemically rearranging the surface film, the most generally used method being to apply a pigmented coating by painting. Painting has largely become, unfortunately, to be regarded as an obligatory final operation to be carried out before products can be got out of the works to bring in income; because of this and in spite of what has been done to emphasize the importance of finish, it does not always receive the consideration and attention that it should. But whatever the finishing method, considerations of surface texture soon enter into the picture. With a very fine texture, a high gloss, care needs to be exercised to ensure that it is not overdone; high reflectivity may accentuate forms by creating strong highlights but it can also contribute to a splintering effect through the reflection of too many light sources and so create an effect of *chiaroscuro*. This effect is more in support of a Mosaic order of form in which the parts tend to dominate the whole, rather than in support of a Gestalt order of form in which the whole appears to dominate the parts. A high gloss also reveals surface defects or imperfections and where a measure of these is unavoidable it can be made less obvious by employing a lower gloss finish. Just lowering the tone is not necessarily the whole answer; a high gloss black finish will reveal surface imperfections just as much as will a similar finish of lighter tone, but it should be appreciated that as gloss is lowered the surface texture is becoming relatively more coarse and more likely to retain surface marks and dirt. This may seem to present a dilemma, made more likely by the employment of multi rather than single colour schemes. For many years convention has decreed that machinery and most industrial equipment should be painted in one colour, usually a grey of medium to dark tone, that being regarded as a sensible choice because it did not readily show surface marks and dirt. Nowadays, for aesthetic reasons in relieving large masses, introducing visual interest and variety and for ergonomic reasons in improving perceptibility in work areas and in indicating static and dynamic parts, etc., two-colour schemes are more common. One of the colours is usually medium to light in tonality and not all who have tried such schemes have managed to cope successfully with the problem of the lighter colour, its being possibly susceptible to revealing surface marking. But then not all have appreciated that such a possibility is due as much to surface texture as to tonality. A colour of medium or even dark tonality and of low gloss is no less liable in this respect than is one of light tonality but with a higher gloss. Experience with one colour, B.S.9–094, a light grey with a tonality of approximately 50 per cent and of 'eggshell' texture which is approximately 50 per cent gloss, showed that it had a tendency to hold and reveal surface marks and would not prove to be suitable, although aesthetically and ergonomically it was entirely suitable. The problem was solved by raising the percentage of gloss to 60, but keeping the tonality as it was, and the scheme has been successfully in use for some years. This simple example highlights the need of some tolerance and patience in dealing with paint finishes, for in spite of the efforts of the finishing industry to introduce a greater measure of reliable and finite specification there is still a large measure of subjectivity on both the user's and supplier's sides. Suppliers have considerable problems in blending constituents to produce a materially acceptable finish and at the same time to achieve desired visual and tactile qualities and a live contact with a paint producer is likely to prove most valuable.

The final selection of an easy-to-maintan surface of the desired hue, which is not too dead to kill the form nor too lively to advertise every

112 THE AESTHETICS OF ENGINEERING DESIGN

defect, is a matter of securing the right balance of tone and gloss, which might be arrived at straight away, but usually after one or two experimental coats. The situation will be helped considerably by the elimination of vague, arbitrary and subjective descriptions such as 'something nearer to matt than eggshell' or 'medium to high gloss' and by the policy of specifying textural quality as a percentage of gloss. Guidance as to the effect of various percentages could come from texture samples obtainable from any good supplier.

Another possibility of serviceable coatings with good visual and tactile qualities is in the employment of thin coatings of various thermoplastic materials which offer a durable finish with a useful range of colours. Unless the equipment for aerating the powdered material for deposition on to the heated parts is self-owned the parts must be sent to

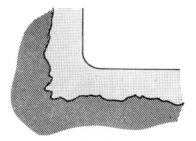

Fig. 6.23

an outside processor, who would, of course, be able to bring to the job a great deal of experience and specialized 'know-how'. Provided that the mechanics of handling, transporting and availability are taken into account this should not introduce any great difficulty and these finishes represent a very useful extension of aesthetic possibility in engineering. Most of the materials used have the virtue of levelling-out and not reproducing faithfully the texture of the surface to which they are applied, Fig. 6.23. This can reduce surface finishing and preparation to the minimum. Some materials, like nylon, also have the virtue of bonding to and not merely encapsulating the part being coated and are especially suitable where heavy impact or abrasion is anticipated and where local piercing of an unbonded coating would be the prelude to complete stripping. The materials being thermoplastic have a natural limitation through their vulnerability to hot metal swarf, but provided that the circumstances of their use are properly examined these coatings can be very useful.

It is, of course, not always necessary, nor desirable, to apply a finish and most natural metal surfaces have good aesthetic quality, although few can be maintained in optimum condition without becoming oxidized. Inoxidizable materials and electro-deposited finishes are therefore used fairly extensively where the use and maintenance of the product does not ensure that an unplated surface will be kept cleaned and oiled, or protected in some other way. The general qualities of most metal film coatings are well known, but it does not always seem to be appreciated how much these finishes can vary in their warmness or their coolness. There are, however, many opportunities for subtlety in the chemistry of metal plating and increased discrimination in specification could extend their aesthetic possibilities. A factor to be considered with metal finishes, as with paint finishes, is the degree of gloss and what might start out as an attractive, even satin texture can soon turn into an uneven textured, locally-polished surface reveailing oil or work stains. Sensible specification suited to the circumstances is essential and care must be taken to avoid selection only upon mechanical merit where visual merit should also be considered.

In the last connection, and indeed in any connection with finishing, it is general practice for engineering design offices to have on file the physical and chemical properties of most generally-used materials. They should also have on file data and samples, rendering the visual and tactile qualities less subject to speculation and to subjective evaluation. There are, for example, at least five different standard finishes for stainless steel, designated broadly as Mill and Polished and incorporating different treatments by abrasives, grinding, annealing, hot or cold rolling and buffing.

A file of actual samples should be an essential part of an engineering designer's reference system; what would we think of a dentist who could only supply chalk-white or oyster-grey false teeth?

In addition to the aesthetic possibilities of steel, colour anodizing of aluminium also offers some very attractive and durable surface finishes. Anodized effects should not be mentally associated only with the often unfortunately trashy and fugitive metallic colours of cheap aluminium ware. It is possible to have good blacks, greys, browns, bronzes and a variety of restrained solid colours through the deposition of light-fast, colour-stable metallic oxides into the pores produced by conventional sulphuric acid anodizing. In deciding upon anodizing, consideration should be given to the particular alloy used, the expected degree of exposure to sunlight or artificial ultra-violet light and to corrosive atmosphere. As with other forms of finishing, the advice and the experience of experts should be sought and will usually be found to be of great value.

Where etched, anodized, plated and unplated engraved metal plates are likely to be used together it is wise to establish acceptable standards of background texture, which can vary from a high polish to a dead mattness. Each standard should be given a name or a value to remove the hazard of different interpretations of what is etched, matt, satin, brushed, bright or polished and, if possible, check samples should be distributed to all concerned. Unless such steps are taken, the maintenance of consistency can prove to be most difficult, particularly where more than one supplier is concerned. Unless some control of visual effect is exercised, one piece of machinery or equipment may exhibit a sufficiently variegated array of control panel or instruction plate finishes to suggest that the design has been of an haphazard nature and the corporate strength of a range of products could be seriously weakened by inconsistency of this kind.

aesthetic aims and the influence of fashion

In any consideration of an overall aesthetic for an engineering product the truth that it should be generated by the facts of a particular situation holds good and skill lies most in recognizing what the facts really are.

Perhaps the most important fact is that the normally-accepted production methods are common to most engineering activities, which naturally imposes similar visual effects of materials and forming processes upon them. But accepting that the aesthetic of engineering production is the first thing to master, it is still possible to temper it with a product-quality, an aesthetic arising out of other facts such as the nature and purpose of the product. For example, machines for the physical removal of metal by another piece of metal, ceramic or a gem, are subject to very similar manufacturing influences in terms of materials and production processes as are machines for removing metal by electro-chemical means. There are, however, sufficient differences in the technological requirements of each basic method and ancillary equipments to produce a justifiable difference in the aesthetic quality of each group. This essential difference is, however, due to the individual facts of each situation rather than to any broader generalizations. It is, for example, incorrect to think of anything so specific as a 'machine tool aesthetic', when there are many kinds of process machines made by similar methods and even using many of the mechanical and structural elements of machine tools. That there can be different aesthetic qualities between different kinds of machine tools makes any concept of a special class aesthetic even more incorrect; it is such thinking that can lead a designer into a blind alley. It is more realistic to think of an aesthetic proper to a particular combination of circumstances rather than to any broad classification of purpose. By so doing, achievement is kept alive and purposeful, and conventional, restrictive stereotyping is discouraged. Another consideration to which newcomers to aesthetics appear to attach importance concerns the extent to which current fashion or stylistic idiom should be consciously incorporated. It is perhaps a prime reason for a great deal of effort proving to be abortive, not because what is done is necessarily wrong, but because the reasons for doing it are wrong. Fashion means no more than a prevailing custom or conventional usage, a reflection of what is currently acceptable to the majority or to an influential minority. With products it must, as already stated, include features especially acceptable in making as well as the acceptability of the concept of the function of the product and of the form it should take. Unless the aesthetically-purist stand can be afforded, which is usually very unlikely, fashionability can be interpreted as acceptability, upon which basis desirable and undesirable qualities are more likely to be correctly assessed.

Products having a short service-life should clearly be geared more closely to the habits and customs of the immediate era than should products with a long service-life. They should thus not only strongly reflect current technologies and manufacturing techniques, but also operational and social preferences which might well be influenced by quite ephemeral qualities arising from other activities of the time. Some engineering products, notably those in the field or short or medium-life consumer durables, belong to this category and it is quite right and proper that their design should integrate them as fully as possible with the current pattern of user habits and customs. A completely non-fashionable aesthetic, generally characterized as a 'timeless quality', may satisfy purists but it cannot be as commercially successful as would an aesthetic suited more completely to the immediate requirement. Capital goods generally and consumer goods with a longer planned service-life need not, indeed should not, reflect so much of the current scene, if they did they would go out of fashion too quickly, while they still had useful years of life left in them.

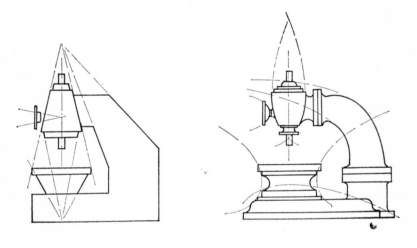

Fig. 6.24

The creation of a strongly marked product-quality need have little to do with fashion although the result may, quite fortuitously, be fashionable; the motive in creating it should be principally to increase the usefulness of the product by indicating clearly its function or purpose. The aim may be psychological, to suggest some emotive quality such as strength, lightness, accuracy or dependability, or it may be linked more to the mechanics of perception by containing visual interest and concentrating it upon certain areas or functions, Fig. 6.24a, and not diffusing it as in Fig. 6.24b.

There should thus be no problem about how consciously fashionable or stylistic a design should be, for the facts of a situation should be termine the solution, as with all considerations of design. Those who cannot properly recognize the facts of a situation should not be practising as designers. Naturally, even where there may be only a low content of current operational and social preferences there may well be a strong reflection of current economical manufacturing methods, which in fact contributes strongly to the contemporary style of many capital goods. As technology and industrial techniques deriving from it change, so we must expect this influence, nothing more than a long-term fashion influence, to change. There should be no problem at all about the fashion content of engineering design, for no human activity can but help reflect something of the time during which it is carried on; if it fails to do this there must be something wrong with it. But design which reflects too strongly personal idiosyncracies, the views of a designer convinced that he is the only one in step and that the rest of mankind is out of step, richly deserves the fate which must ultimately befall it, which is certain and merciful oblivion.

determination of product quality

Reference elsewhere in this book to preferred visual arrangements and to aesthetic manoeuvre is concerned principally with certain aspects of design and not with design in a broader sense, such as the determination of a rationale or an overall product quality. Of course, the development of a more sure handling of minor aspects, the acquisition of tactical skill, must automatically generate a sounder and bolder approach to major aspects of design, the strategy. But the determination of a general product quality must be of especial interest to many who already have some bases for a design character, be they biased toward structural requirements, mechanical function or manufacturing convenience. Such bases, can, however, easily become individually too dominant without producing an optimum product quality, and it is unwise to be in any way dogmatic about this matter as compromise of different requirements is implicit in good design. A more fruitful course is to preserve an open mind which ensures the recognition of all factors that should be considered and which will, in each particular circumstance, lead to an optimum balance of those factors.

Some products may be automatically characterized by their purpose, by the functional intention, and the efforts of the makers may be so completely identified with the requirements of the user that a strong commercial approach, biased towards the interests of the manufacturer, can be adopted without prejudice to product usefulness or to the user's interests. But design situations are rarely so clearly or satisfactorily defined and it is usually necessary to emphasize certain characteristics to indicate the purpose of the product and to encourage best use of it, while avoiding counteraction between the qualities so emphasized and bearing in mind all aspects of production so that manufacture and promotion remains an economic and worthwhile activity. This is by no means easy, but possible conflict between undue attention to the manufacturer's interests and insufficient attention to the user's interests is best avoided by an objective and liberal assessment of the product and of the whole exercise of creating it. This is where designers enjoy some advantage over those whose interests are more deeply committed perhaps, but over a narrow front. Where the major concern is with the return upon an investment or in the maintenance of a position in the face of competition, fair and unfair, it is certainly desirable that an open-minded and self-critical attitude be maintained, but it can reveal the need of changes in established manufacturing and marketing policies and techniques. Any change in what exists naturally introduces an element of speculation which may be for better or for worse, and it is almost certainly easier for designers to maintain an entirely objective approach than it is for others more heavily committed. The amortization of expenditure involved in tooling and in various forms of sub-contracting need not present problems and can lead to increased profitability, but it can also lead to the perpetuation of designs and product qualities that are themselves aggravating and prolonging an unprofitable situation. It is impossible to absolve designers entirely from responsibility for such situations, but it is possible to make every effort to ensure that they do not arise.

In weighing manufacturing convenience against product performance and usefulness the balance is naturally tipped in favour of the latter, in that they contribute to better saleability and to a healthier marketing situation. It is thus usually preferable to base the rationale of a design upon qualities most likely to ensure optimum indication of purpose and favour maximum usefulness, unless the situation is such a highly competitive one that very small margins of production cost could significantly affect success. But even where full recognition of manufacturing convenience and economy must be made, it is still possible to achieve expression of other qualities, as the attractive, efficient and highly competitive machine tool referred to on pages 20–23 shows.

The selection of an overall visual characteristic is not easy and should rarely be confined to one quality. For example, a product may need to be structurally very strong, mechanically very efficient and functionally very accurate, but any one of those qualities could be visually emphasized to the detriment of the others. It is therefore generally necessary to synthesize several principal qualities in such a way that any one of them can be clearly read when that quality is being looked for. The best hope for the realization of this is the abandonment by engineering designers of very strong personal idioms and idiosyncracies and the adoption of a truly objective approach to visual matters. This by no means rules out the possibility of a personal style or 'handwriting', but it does ensure that any visual statement made will be understood by, and accepted by, a majority rather than a minority of people. Beyond indicating a necessary mental attitude, and without an actual case to consider, it is not possible to be specific in this matter of the determination of an overall design characteristic; some generalizations may however be helpful. Many purely engineering requirements can be met equally well by alternative physical arrangements, the visual manifestation of one of which will usually comply better with a desired aesthetic aim than will the others. For example, where extreme precision and accuracy are dependent upon great strength and stiffness of structure, it is possible for the visual expression of one quality to suggest the absence of the others. The expression of one quality, say mechanical efficiency, could be in conflict with the visual expression of structural strength. Secondary forms such as conveniently exposed motors, external gear boxes, electrical boxes and swarf and dust collecting gear must inevitably compete with the principal form, which is generally the main structure. Add ventilation openings and a garnish of hardware for air, oil and electrical services and the possibility of a strong, unified principal form being dominant begins to recede. To impose poor accessibility and bad ventilation through concealment, merely to achieve visual simplicity, or some stylistic aim, is bad engineering design, but there is more than one order of form and a satisfactory result is usually a matter of appropriate application. Where it is clear that the whole cannot condition the parts without prejudice to physical qualities the Gestalt order of form is not indicated; the Mosaic order of form, where the parts condition the whole, may however be quite appropriate. Real awareness of matters of form requires greater study than a perusal of the glossary of terms in this book, but it is quite essential in engineering design that visual values be accorded proper importance. The achievement of mechanical and structural aims alone cannot ensure that vital spark of life, an adequate product quality, to the achievement of which the senses and intangible feelings are essential and must be rated as important as the most materialistic and technical contributions. A disability suffered by most engineers and engineering designers, traceable to inadequate design education and to specialization, is a bias towards structures, mechanisms, services or some other single aspect of engineering design, which makes the essential synthesizing of visual characteristics difficult for them. This disability has indicated an area in which industrial design specialists have been able to make an indisputable contribution, but, as implied throughout this book, this specialist contribution is limited because the proposals of engineers and engineering designers must be 'frozen' at an early stage of project development to enable parallel development to proceed and this can make subsequent synthesizing most difficult, even for a visual specialist. The only really satisfactory solution lies in a reassessment of engineering design education and the making good of current glaring omissions from it. Until this happens, the existing situation and the sincere beliefs of those who see the reduction to multi-specialization of what should be a complete, unified activity, can only be seen as expediency and illusion.

glossary of aesthetic terms

AIRY   Applied to areas of forms or to compositions of forms where there is not much to engage the attention. It can describe a visual lightness or emptiness and may often be prefixed by the word 'too'.

ANATOMICAL   Applies to the basic, physical arrangement of form, to the skeleton rather than to the flesh or the skin. Getting the anatomy right should be a basic aesthetic aim because overdependence upon cosmetic treatment alone results in dressing-up, facelifting, superficial packaging or boxing-in. A greater awareness of sound aesthetic quality would result in the anatomy of engineering design being healthier than it often is and would reduce the necessity for so many major operations. See FORM.

BALANCE   This does not necessarily mean symmetry, a popular misconception. Asymmetry can be visually balanced and be less restricting and stultifying than symmetry can often be, particularly in the context of the presentation of controls and information where symmetry might be quite inconsistent with proper sequential arrangement. In arriving at good visual balance there is a distinct connection with experience of physical balance and a subconscious evaluation of weights and leverages to produce balancing moments: an unconscious reference back to a knowledge of mechanics. Symmetry is not, of course, without merit and can sometimes be the best way of securing visual balance, but it is not the only way and often results in forms failing to achieve full aesthetic potential.

CHIAROSCURO   This is another term applying to light and shade, or to black and white combinations especially those reminiscent of chess-board markings, or to any area that is visually 'busy' with marked tonal gradations. In the engineering design context it could apply to the surface quality of a form that has become embellished with a number of secondary forms at the expense of its own identity. See FORM – Mosaic order of.

COLLAGE   A term invented by a Continental school of painters, referring to the creation of an image by pasting on to drawn or painted areas a variety of disparate objects such as bus-tickets, bottle-tops, news-cuttings and photographs. In the engineering design context it could describe an inappropriate application of graphic elements, possibly associated with control panels or information areas.

COLOUR   Aesthetically, this does not necessarily signify hue or chromatic conditions and it can refer to the overall visual value or richness. Colourless may not necessarily mean the dilution or absence of hue, but a general weakness of aesthetic quality, visual sterility or barrenness. Colourful can describe visual richness and abundance of aesthetic quality, but does not necessarily imply the presence of strong hue.

CONCEPT   An idea, mental plan or general notion. An abstraction, without necessarily any image of a possible visual manifestation.

COSMETIC   Meaning superficial or concerned with the surface and could apply to surface finish and to all applied features such as name and information plates, controls and fastenings. Not all of these things may be functionally superficial, although aesthetically they are applied parts of the exterior which are seen and which may influence, but do not determine, the basic physical arrangement of the form. See FORM.

DIRECTION   In its aesthetic context this has nothing to do with movement; completely static forms can have direction. It has more to do with a visual emphasis, preferably reflecting some functional quality, which makes the form more readily identifiable and its purpose understood. Direction can thus be seen as an aid to perception and where the directions of several forms converge or diverge the result is to concentrate or to dissipate attention, the former being more likely to be a deliberate aim than the latter.

FAT   Does not necessarily mean physically full, but rich and full of visual interest. See IMPASTO.

FIGURE   Describes any mark and may have arisen from the fact that most early paintings, *scraffiti* and carvings were representations of human beings or animals. Most marks can be figures only in the aesthetic context, excepting numerals which are figures in the numerical sense as well and to avoid confusion it is better to refer to numbers as numerals and not as figures.

FORM   Means shape, configuration, arrangement of parts, organization and system of relationships. It usually implies something three-dimensional, shape being applied more to two-dimensional things. There is the manifestation of internal form, the structure, and that of the external form, the visible shape; these are analogous to anatomical and cosmetic qualities.

The Gestalt Theory of form is where the whole appears to dominate the parts and is applicable to most man-made forms, which are usually designed to serve a function. There is also a Mosaic order of form, characterized by sharply defined parts and associated more with biological forms.

GROUND   The background against which any figure, or mark, is seen.

HARD EDGE   A term perhaps more appropriate to two-dimensional graphic effect, but applicable to sharp, clear-cut, three-dimensional form, particularly where individual forms in a composition require to be emphasized rather than assimilated into the group. See FORM – Mosaic order of.

HEAVY   Does not refer only to weight. It means a combination of qualities or the inappropriate placing of a form or a mark making it appear too heavy within the context of its use. See WEIGHT.

IMPASTO   This is a painting term meaning laying it on thickly, but it could be used to describe an overrich visual effect.

MASS   In the aesthetic context this applies to area and volume and not only to weight.

MOVEMENT   Movement is an expression of direction deliberately emphasized. It should be associated sensibly with the function of the form and it can be applied to forms that do not themselves move but which are strongly associated with movement. This could be where a mechanical function is concealed but where the results of it emerge and the concealing form should be given visual movement to direct attention to the point of emergence. Apt suggestion of movement could contribute to the safety of some forms.

ORGANIZATION   This has the normal meaning although aesthetically it may be applied to the arrangement of intangibles; an analogy may be drawn between organized and unorganized forms and groups of people in similar states. It implies an harmonious arrangement of parts and it always implies the application of conscious effort.

PERCEPT   The object or product of perception. As the former it need not already have a material existence, but some possible visual state can be imagined. It is a concept brought closer to actuality.

PLASTIC   In the aesthetic context this has nothing at all to do with synthetic materials nor with any ability to bend or flow. It signifies an actuality, a totality or the visual sum-total of a form. It is the product of visual weight, form, colour and texture, including any visible defects. If a form could say 'Here I am, take me as you find me' it would be referring to its total plastic quality. Plastic quality implies a three-dimensional state, an occupation of space and not merely a representation of space as in a perspective drawing. But it is important to remember that it means more than just three-dimensional, that it is a comprehensive evaluation of form, colour, texture and visible quality.

RANGING   This describes the alignment of type matter, or a number of forms, in relation to a datum which could be their visual or geometric centres or one edge.

READING   Normally associated only with the perception of words, although it applies equally well to the perception of any kind of visual display. Displays that are visually well-organized read well, like concise wording, while less-well-organized displays are confusing and difficult to understand, like obscure or ambiguous wording.

TENSION   A noun connected with balance and arrangement, with placing and relationships and susceptible to analogy with actual physical links and their elastic limits. Objective analysis and evaluation of tension will usually coincide with intuitive judgment or 'feeling'.

THIN   Does not necessarily mean physically slim, but low in visual interest. In the aesthetic context it means transparent, shallow and lacking in richness.

UNITY   Meaning a oneness, a coherence of the parts forming the whole in the case of a single object, or a family-likeness or corporate style in the case of a number of different objects. Unity involves the visual characteristics of the parts as well as their relationship and the fundamental basis of unity is similarity.

WEIGHT   This refers to the visual weight of a form which is determined not only by its colour but by a product of the characteristics of its surface which includes texture and colour and of characteristics of its form.

Forms with certain characteristics can appear to be lighter than other forms although they may be physically larger. Large radii, sagging as opposed to stiff curves, thick as opposed to thin overlaid forms, coarse as opposed to fine texture; all contribute to increasing visual weight. Crisp radii, stiff curves, thin overlaid forms and fine texture contribute to visually lighter forms. Actual weight is not necessarily a corollary of stiffness or strength; forms can still reflect these qualities while being clean, simple in mass and not visually heavy.

index

Aesthetic aims and influence of fashion, 114–17
Aesthetic manoeuvre, 74–91; balance, 87–90; clarity of expression, 85–7; clarity of form, 82–5; positioning and relationship, 79–82; symmetric and asymmetric form, 87, 90–1; visual arrangements, 79–82
Agricola, books on engineering, 2
Alpha rhythms, 31
Aluminium, anodized, 113–14
Anatomical form, 52, 53, 121
Archimedes, 2: Archimedian spiral, 62–4
Architecture and engineering, 3
Asymmetry, see Symmetry
Attenuation, 68–70

Balance, 87–90, 121; and layout, 100–4; axis or pivot point, 88–9; static and dynamic, 87, 90, 104
Basic visual elements, 39–44: abstractions without weight or movement, 39; form, 41–2; curved element, 43–4; horizontal element, 42–3; oblique element, 43; relationships of shapes and forms constructed from basic elements, 44; variation of figure on ground, 39; vertical element, 43
British Standards Institution:
BS 2660 (colour range), 109;
BS 3641 (*Symbols for machine tool indicator plates*), 107–8;
Mechanical Engineering Standards Committee, 108

Capital lettering, 93–4
Chiaroscuro, 110, 121
Chroma, definition of, 109
Clarity of expression, 85–7
Clarity of form, 82–5; and structural validity, 84–5
Clarity of presentation, 93
Coatings, 110–13; thermoplastic, 112–13
Colour, 104, 108–14, 122; anodized aluminium, 113–14; *chiaroscuro*, 110; coatings, 110–13; file of samples, 113; finishes, 110–14; line, value and chroma, 109; natural metal surfaces, 113; range, 109; selection, 109–14
Commercial value of aesthetics, 16–24: examples, 17–23; value-analysis, 24; visual thinking, 16–17
Composition, 51–3: anatomical form, 52, 53; qualities required, 52
Cosmetic quality, 51, 53, 54, 101, 108, 122
Curved element, 43–4

D.I.N. Colour Dictionary, 110
Direction, 70–8, 122: basic considerations of direction in forms, 70–3; creation of sense of direction, 73–8; visual organization, 76–7
Display nicety, 104–6
Dynamic balance, 87, 90, 104

Elegance, 6
Engineering as art, 2–7: aesthetic quality, 5–7; elegance, 6; objectivity, 2–3

128 INDEX

Ergonomics, 15–16; differences from aesthetics, 15–16

Fibonacci, 62
Figure, 39, 41, 122: on ground, 39, 79–82
Finish, 92: colour finishes, 110–14
Folio typeface, 96
Form, 41–2, 51–78, 122–3: clarity, 82–5; composition, 51–3; direction, 70–8; harmonic relationship, 64–6; internal and external form, 51; proportion, 58–64; secondary forms, 66–70; surface, 53–5; symmetric and asymmetric, 87, 90–1
Function of aesthetics in engineering design, 11–29: aesthetic of shape, form and arrangement, 12; commercial value of aesthetics, 16–24; ergonomics, 15–16; indication of function and purpose, 13; material aesthetics, 12; opportunities to achieve aesthetic quality, 12–14; optimization, 24–9; priority of engineering function, 11–12; rationalization, 24–9; satisfaction of human requirements, 14–15; simplification, 25–7; variety of materials, 12; variety of surface treatment, 12; visual thinking, 14
Futura typeface, 96

Gestalt theory of form, 110, 120, 123
Gill typeface, 96
Goethe, 3
Golden Mean System, 60–2
Graphic display, 92
Greek engineers and geometricians, 2, 62
Grotesque typeface, 96
Ground, 39, 41, 123; figure on ground, 39, 79–82
Gutenberg, Johannes, 96

Harmonic relationship, 64–6; and formal systems of proportion, 64–5; and unification, 55; subdivision, 64–6
History of engineering, 2
Horizontal element, 42–3
Hue, definition of, 109
Human aspects of engineering design, 3–4

Illusory optical effects, 36–7
Irrationality, 30–1, 33–9, 47–8, 84: apparent structural improbability, 33; examples, 35–8; identification, 33; illusory optical effects, 36–7
Italic lettering, 95

Layout and balance, 100–4: brevity and clarity, 100; symmetry and asymmetry, 101–4
Leonardo da Vinci, 4–5, 28, 46, 95: as artist-engineer, 4; notebooks, 4
Lethbridge, T. C., 63
Letters, 92–9: choice of typeface, 93–9; clarity of representation, 93; selection and arrangement of content, 99; size, 106–8
Logarithmic spiral, 62

Material aesthetic, 12
Maudslay, Henry, 5
Metal surfaces, 113–14
Mosaic order of form, 110, 120
M.T.I.R.A. Research Report No. 10, 108
Munsell system, 109

Nasmyth, James, 5
Nicety of display, 104–6
Numerals, 98–9: size, 106–8

Oblique element, 43
Optimization, 24–9

Perception, 14, 30–50, 65: Alpha rhythms, 31; basic visual elements, 39–44; designer's imposition of interpretations, 32–3; irrationality, 30–1, 33–9, 47–8; self-deception, 32; sense-impressions, 46–8; sense-memory, 46, 48–50; visual competition, 44–6; visual priority, 46
Persian engineers, 2
Pick, Frank, 93
Ponzo illusion, 36–7
Positioning and relationship, 79–82
Product-quality, 116: determination, 117–20; selection of overall visual characteristics, 119; synthesis of several principal qualities, 119
Proportion, 58–64: and harmonic relationship, 64; formal systems, 58–64; orthographic view, 58–60; suitable ratio within context, 58

INDEX

Rationalization, 24–9
Roman lettering, 96–7
Root Five Rectangle system, 60, 61, 64

Sans-serif faces, 96–8
Secondary forms, 66–70, 119:
    attenuation, 68–70
Sense-impressions, 46–8, 67, 85
Sense-memory, 46, 48–50
Serifs, 95–8; sans-serif faces, 96–8;
    slab serifs, 97
Shun, Chinese Emperor, 2
Simplification, 25–7
Size of displayed letters or numerals,
    106–8: degree of legibility, 107–8
Slab serifs, 97
Static balance, 87
Step in the Dark, A
    (Lethbridge), 63
Structural validity, 84–5
Summation Series, 62
Surface and surface treatment, 12, 53–5,
    92–120: aesthetic aims and influence
    of fashion, 114–17; colour, 104,
    108–14; finish, 92; graphic display,
    104–6; layout and balance, 100–4;
    letters, 92–9; nicety of display, 104–6;
    numerals, 98–9; product-quality,
    119–20; size, 106–8
*Symbols for machine tool indicator
    plates* (BS 3641), 107–8
Symmetry and asymmetry, 121; of
    form, 87, 90–1; of layout, 101–4

Thermoplastic coatings, 112–13
Typeface, choice of, 93–9: capitals,
    93–4; italics, 95; Roman lettering,
    96–7; serifs, 95–8; upper and lower
    case, 93–4

Univers typeface, 96
Upper and lower case lettering, 93–4

Value analysis, 24
Value of colour, 109
Venus typeface, 96
Vertical element, 43
Visual arrangements, 79–82
Visual axis or pivot point, 88, 89
Visual balance, 87–90
    See also Balance
Visual competition, 44–6
Visual organization, 76–7, 123
Visual priority, 46
Visual thinking, 14, 16–17
Visual unity, 55–8, 65, 66, 124: and
    relationship of visual characteristics,
    56–7; connection between unification
    and harmonic relationship, 55;
    origin, 55–6

Whitehead, Professor A. N., definition
    of style, 24

Yu, Chinese Emperor, 2